EAI/Springer Innovations in Communication and Computing

Series Editor

Imrich Chlamtac, European Alliance for Innovation, Ghent, Belgium

The impact of information technologies is creating a new world yet not fully understood. The extent and speed of economic, life style and social changes already perceived in everyday life is hard to estimate without understanding the technological driving forces behind it. This series presents contributed volumes featuring the latest research and development in the various information engineering technologies that play a key role in this process. The range of topics, focusing primarily on communications and computing engineering include, but are not limited to, wireless networks; mobile communication; design and learning; gaming; interaction; e-health and pervasive healthcare; energy management; smart grids; internet of things; cognitive radio networks; computation; cloud computing; ubiquitous connectivity, and in mode general smart living, smart cities, Internet of Things and more. The series publishes a combination of expanded papers selected from hosted and sponsored European Alliance for Innovation (EAI) conferences that present cutting edge, global research as well as provide new perspectives on traditional related engineering fields. This content, complemented with open calls for contribution of book titles and individual chapters, together maintain Springer's and EAI's high standards of academic excellence. The audience for the books consists of researchers, industry professionals, advanced level students as well as practitioners in related fields of activity include information and communication specialists, security experts, economists, urban planners, doctors, and in general representatives in all those walks of life affected ad contributing to the information revolution.

Indexing: This series is indexed in Scopus, Ei Compendex, and zbMATH.

About EAI - EAI is a grassroots member organization initiated through cooperation between businesses, public, private and government organizations to address the global challenges of Europe's future competitiveness and link the European Research community with its counterparts around the globe. EAI reaches out to hundreds of thousands of individual subscribers on all continents and collaborates with an institutional member base including Fortune 500 companies, government organizations, and educational institutions, provide a free research and innovation platform. Through its open free membership model EAI promotes a new research and innovation culture based on collaboration, connectivity and recognition of excellence by community.

Manolo Dulva Hina • Seyedali Mirjalili •
Amar Ramdane-Cherif • Rafik Zitouni
Editors

Future Research Directions in Computational Intelligence

Selected Papers from the 3rd EAI International Conference on Computational Intelligence and Communication

Springer

EAI
RESEARCH MEETS INNOVATION

Editors
Manolo Dulva Hina
ECE Research Center, Omnes Education
ECE Engineering School
Paris, France

Seyedali Mirjalili
Centre for AI Research and Optimisation
Torrens University Australia
Brisbane, QLD, Australia

Amar Ramdane-Cherif
LISV laboratory
University of Versailles - Paris-Saclay
Vélizy, France

Rafik Zitouni
ICS - 5G & 6G Innovation Centre
University of Surrey
Guildford, Surrey, UK

ISSN 2522-8595 ISSN 2522-8609 (electronic)
EAI/Springer Innovations in Communication and Computing
ISBN 978-3-031-34458-9 ISBN 978-3-031-34459-6 (eBook)
https://doi.org/10.1007/978-3-031-34459-6

This Springer imprint is published by the registered company Springer Nature Switzerland AG
The registered company address is: Gewerbestrasse 11, 6330 Cham, Switzerland

Paper in this product is recyclable.

Preface

We are delighted to introduce the proceedings of the third edition of the European Alliance for Innovation (EAI) International Conference on Computational Intelligence and Communications (CICom 2022). This conference has brought researchers, developers, and practitioners around the world who are leveraging, developing, and applying computational intelligence and advances in communications for a better today and a smarter tomorrow. The theme of CICom 2022 was "Trustworthy Artificial Intelligence: Reliable computational intelligence solutions".

The technical programme of CICom 2022 consisted of eight full papers. The conference tracks were Track 1: Computational Intelligence in Automation, Control, and Intelligent Transportation System; Track 2: Computational Intelligence on Big Data, Internet of Things, and Smart Cities; Track 3: Computational Intelligence on Wireless Communication Systems and Cyber Security; and Track 4: Computational Intelligence on Human-Machine Interfaces, and Image and Pattern Recognition. Aside from the high-quality technical paper presentations, the technical programme also featured two keynote speeches. The two keynote speakers were Dr Sofiane Abbar, Machine Learning Engineer at Meta (Facebook/WhatsApp), and Dr Ali Safaa Sadiq from Nottingham Trent University, United Kingdom.

Coordination with the steering chair, Imrich Chlamtac, and EAI conference manager, Kristina Havlickova, was essential for the success of the conference. We sincerely appreciate EAI's constant support and guidance. It was also a great pleasure to work with such an excellent organizing committee team for their hard work in organizing and supporting the conference, in particular, the Technical Program Committee, led by Dr Manolo Dulva Hina, Dr Amar Ramdane-Cherif, and Dr Yassine Meraihi who have completed the peer-review process of technical papers and made a high-quality technical programme. We are also grateful to Dr Piotr Kuwalek, Dr Aakash Soni, Dr Guilherme Medeiros Machado, and Dr Ravi Tomar for their support during the conduct of the conference and all the authors who submitted their papers to the CICom 2022 conference.

We strongly believe that CICom conference provides a good forum for all researcher, developers, and practitioners to discuss all recent advances, challenges and opportunities, and perspectives relevant to computational intelligence and communications. We also expect that the future CICom conferences will be as successful and stimulating as indicated by the contributions presented in this volume.

Paris, France — Manolo Dulva Hina
Brisbane, QLD, Australia — Seyedali Mirjalili
Vélizy, France — Amar Ramdane-Cherif
Guildford, Surrey, UK — Rafik Zitouni

CICom 2022 Organization

Steering Committee

Imrich Chlamtac	Bruno Kessler Professor, University of Trento, Italy
Manolo Dulva Hina	ECE Paris Engineering School, France

Organizing Committee

General Chair

Manolo Dulva Hina	ECE Paris Engineering School, France

General Co-Chairs

Seyedali Mirjalili	Torrens University, Australia
Rafik Zitouni	University of Surrey, UK
Amar Ramdane-Cherif	University of Versailles – Paris Saclay, France

TPC Chair and Co-Chair

Manolo Dulva Hina	ECE Paris Engineering School, France
Amar Ramdane-Cherif	University of Versailles – Paris Saclay, France
Seyedali Mirjalili	Torrens University, Australia
Jaouhar Fattahi	Université Laval, Canada
Piotr Kuwalek	Poznan University of Technology, Poland
Yassine Meraihi	Université de Boumerdes, Algeria

Sponsorship and Exhibit Chair

Hongyu Guan	University of Versailles – Paris Saclay, France

Local Chair

Amar Ramdane-Cherif	University of Versailles – Paris Saclay, France

Workshops Chair

Yassine Meraihi	Université de Boumerdes, Algeria

Publicity & Social Media Chair

Naila Bouchemal	ECE Paris Engineering School, France

Publications Chair

Rafik Zitouni	University of Surrey, UK
Aditi Sharma	Parul University, India

Web Chair

Ravi Tomar	University of Petroleum and Energy Studies, India

Posters and Ph.D. Track Chair

Samir Brahim Belhaouari	Hamad Bin Khalifa University, Qatar

Panels Chair

Piotr Kuwalek	Poznan University of Technology, Poland
Abderrahmane Maaradji	ECE Paris Engineering School, France

Demos Chair

Sebastien Dourlens	University of Versailles – Paris Saclay, France

Tutorials Chairs

Jaouhar Fattahi	Université Laval, Canada
Moeiz Miraoui	Umm Al-Qura University, Saudi Arabia

Technical Programme Committee

Manolo Dulva Hina	ECE Paris Engineering School, France
Jaouhar Fattahi	Université Laval, Canada
Yassine Meraihi	Université de Boumerdes, Algeria
Piotr Kuwalek	Poznan University of Technology, Poland
Amar Ramdane-Cherif	University of Versailles – Paris, Saclay, France
Hongyu Guan	University of Versailles – Paris, Saclay, France
Sébastien Dourlens	University of Versailles – Paris, Saclay, France
Naila Bouchemal	ECE Paris Engineering School, France
Aakash Soni	ECE Paris Engineering School, France
Abderrahmane Maaradji	ECE Paris Engineering School, France
Jae Yun Jun Kim	ECE Paris Engineering School, France
Ali Awde	Collège St-Foy, Canada
Seyedali Mirjalili	Torrens University, Australia
Ravi Tomar	University of Petroleum and Energy Studies, India
Naresh Kumar	Quantum University, India
Aditi Sharma	Parul University, India
Moeiz Miraoui	Umm Al-Qura University, Saudi Arabia
Rolou Lyn Maata	Gulf College, Oman
Samir Brahim Belhaouari	Hamad Bin Khalifa University, Qatar
Rafik Zitouni	University of Surrey, UK

Contents

About the Editors

Manolo Dulva Hina is currently an associate professor at ECE Research Center, ECE Engineering School (ECE Ecole d'Ingénieurs), Paris, France. He obtained his Ph.D. in Computer Science from the University of Versailles – Paris Saclay in Versailles, France, in 2011, and his Ph.D. in Engineering (Applied Research) from Université du Québec, Ecole de technologie supérieure, Montréal, QC, Canada, in 2010. He has more than 20 years of teaching experience in Computer Science and Engineering, having taught in various colleges and universities in the Philippines, Canada, Bahrain, and France. He also served as Dean of the College of Computer Studies in University of Technology Bahrain (formerly AMA International University Bahrain). He has edited several books and published several book chapters on computational intelligence. He has also published several papers in top-tier journals, and in international conferences, in which six of his papers won "Best Paper" awards. He was also invited as Keynote Speaker in several international conferences. He serves as Chair of the annual EAI/Springer International Conference on Computational Intelligence and Communication (CICom). His research interests are in the area of computational intelligence, intelligent transportation system, ambient intelligence, human-machine interactions, and formal specifications.

Seyedali Mirjalili is a professor at Torrens University's Center for Artificial Intelligence Research and Optimization and is internationally recognized for his advances in nature-inspired Artificial Intelligence (AI) techniques. He is the author of more than 500 publications with over 70,000 citations and an H-index of 85. He is one of the most influential AI researchers in the world. From Google Scholar metrics, he is globally the most cited researcher in optimization using AI techniques, which is his main area of expertise. Since 2019, he has been in the list of 1% highly cited researchers and named as one of the most influential researchers in the world by Web of Science. In 2022, *The Australian* newspaper named him a global leader in Artificial Intelligence. He is a senior member of IEEE and is serving as an editor of leading AI journals including *Neurocomputing*, *Applied Soft Computing*,

Advances in Engineering Software, *Computers in Biology and Medicine*, *Healthcare Analytics*, *Applied Intelligence*, and *Decision Analytics*.

Amar Ramdane-Cherif received his Ph.D. from Pierre and Marie Curie University in Paris in 1998. In 2007, he obtained his HDR degree from the University of Versailles. From 2000 to 2007, he was an associate professor at the University de Versailles and worked in PRISM Laboratory. Since 2008, he is a Full Professor at the University of Versailles – Paris Saclay, working in the LISV laboratory. His research interests include: (1) Software Ambient Intelligence – semantic knowledge representation, modelling of ambient environment, multimodal interaction between person, machine and environment, fusion and fission of events, ambient assistance; and (2) Software Architecture – software quality, quality evaluation methods, functional and non-functional measurement of real-time, reactive, and software-embedded systems. As of today, he has edited two books on computational intelligence and communications. He authored ten book chapters, 60 international journals, and about 150 international conference papers. He also has supervised 20 doctoral Ph.D. theses and reviewed some 30 Ph.D. theses. He managed several projects and has been doing several national and international collaborations. He is currently a member of the Council Board of the Graduate School of Computer Science of the Paris-Saclay University.

Rafik Zitouni is a senior research fellow at the University of Surrey and Lead Research Developer at the 5G and 6G Innovation Centers. He is a highly accomplished research engineer and academic with a wealth of experience in computer science. His expertise in new architecture and protocols for 5G New Radio and 6G networks has led to significant contributions to numerous international projects. Previously, he was a lecturer at the ECE Paris Graduate School of Engineering in France and a research associate at the VEDECOM Institute in Versailles, France. He holds a Ph.D. in computer science from the University of Paris-Est and an M.Sc. from the University of Paris 12. He also obtained engineer and magister degrees in Artificial Intelligence and Computer Networks from the universities of Setif and Bejaia in Algeria. Dr Zitouni's research interests include wireless network protocols, software and cognitive radios with artificial intelligence, and software design. He has published numerous research articles in top-tier academic journals, and his expertise has led to invitations to speak at conferences around the world. In addition to his technical research subjects, he has taken a leadership role in organizing research conferences in Africa and Europe. Dr Zitouni has also co-edited several books on new technologies for emerging countries and on computational intelligence.

Part I
Computational Intelligence for All

Chapter 1
Multilingual Context-Aware Chatbots for Multi-domain Test Data Generation

Til Weissflog, Mathias Leibiger, Daniel Fraunholz, and Hartmut Koenig

1.1 Introduction

The development of forensic methods and tools for analyzing chat communication and telephone conversation requires appropriate test data. Currently, such data sets have to be produced manually in most cases. This is cumbersome because it requires a large amount of resources regarding personal, time, and money. Automating this process through generating artificial test data would allow for a faster low-cost generation of a wide range of data sets for multiple forensic domains. In addition, forensic test data sets have to combine both general and domain-specific content. The higher proportion of this data has to be general and irrelevant for search algorithms, i.e., context descriptions or everyday talks. The remaining proportion, in contrast, presents domain-specific or case-related content, e.g., hate speech, organized crime talks, etc., to model the communication of a suspected person over a longer period of time. For the generation of such artificial forensic data sets, *natural language processing* (NLP) methods and data sets are required. Especially, data sets for the evaluation of novel NLP and analysis methods are important to further facilitate the potential of these technologies. Such data sets are, however, also rarely available and restricted to a few languages, especially in the form of conversational texts. In this chapter, we present a chatbot framework

T. Weissflog (✉)
Cyberagentur, Halle, Germany

ZITiS, Communications Department, Munich, Germany
e-mail: Til.Weissflog@zitis.bund.de; weissflog@cyberagentur.de

M. Leibiger · D. Fraunholz · H. Koenig
ZITiS, Communications Department, Munich, Germany
e-mail: Mathias.Leibiger@zitis.bund.de; Daniel.Fraunholz@zitis.bund.de;
Hartmut.Koenig@zitis.bund.de

M. D. Hina et al. (eds.), *Future Research Directions in Computational Intelligence*, EAI/Springer Innovations in Communication and Computing,
https://doi.org/10.1007/978-3-031-34459-6_1

for synthetically generating forensic data sets for chat and audio communication. It applies various NLP methods. Unlike most state-of-the-art chat bots that are human-to-bot systems, our framework is designed as a bot-to-bot one. The bots act autonomously and generate the synthetic data automatically. The synthetic data sets are generated as raw data in ASCII text that is then transformed into defined data types, e.g., messenger and audio formats. As a result, we receive forensic data sets with realistic and domain-specific content that can be used for the evaluation of forensic tools and methods. The remainder of this chapter is structured as follows. In Sect. 1.2, we give an overview of current available NLP data sets. We compare state-of-the-art models for text and audio generation with the focus on suitability for artificial data generation. Section 1.3 introduces our framework and describes the data generation procedure. In Sect. 1.4, we evaluate the quality of the generated data and discuss their customizability and usability for real use cases. Some final remarks conclude the paper.

1.2 Models for Natural Language Generation and Data Sets

1.2.1 Data Sets

There are several large data sets for NLP, but only a few of them can be used to analyze the communication behavior of suspected persons. The so-called *persona-chat* [23] data set does not support domain-specific phrases. Another data set is called *M-Ailabs* [18] and contains spoken audio samples. These audio samples also do not contain domain-specific data. Another example is the "*Enron Email*" data set [5] consisting of 500,000+ e-mails with text and metadata, but it can only be used for a limited number of machine learning tasks. There is no possibility to modify them for a specific domain or changing the style of writing. Furthermore, the format is limited to e-mail addresses. *SMS Spam Collection Data Set* [2] is another set that contains 4,000+ SMS messages, labeled with a spam or ham (genuine e-mail) attribute. It can be used for the training of classification algorithms, but it does not contain long personalized chats for semantic analysis or data extractions. *Spoken Wikipedia Corpora* [12] is an audio data set for German, English, and Dutch. This set contains read Wikipedia articles. It can be applied to different tasks, such as speech-to-text or semantic analyses. It does not allow one to insert domain-specific audio chats or change the voice, which is essential for evaluating forensic tools and algorithms. Also the written text in Wikipedia does typically not reflect real-world conversational text styles or forensic use cases. To sum up, although there are various text and audio data sets, all of them are not customizable for domain-specific use cases. Most of them are not suitable for analyzing long chat histories because they consist only of single messages without a chat context. Furthermore, most of the data sets do not contain metadata that is essential in forensic analyses for finding correlations or for getting information about (partially) encrypted or encoded messages.

1.2.2 Models

Generating synthetic text data requires a generative NLP model. In the past, there have been different kinds of models, such as *Recurrent Neural Networks* (RNNs) and *Gated Recurrent Units* (GRUs). In 2017, Vaswani et al. have introduced transformer architectures [19] that have become the predominant method for NLP tasks meanwhile. The concept of a transformer is to analyze data while considering the data context. For this, the transformer combines neural networks and matrix operations for weighting the values of each input value in a multidimensional vector space. The results are new values with a strong correlation to the context inputs. The models that use this technique for generation tasks are called *generative-pretrained-transformer* (GPT) models. Table 1.1 compares several state-of-the-art GPT-models. The models [4, 14] published by the research group OpenAI mainly differ in the model size. The larger the size of the model, the larger is the amount of needed training data and consequently the total training cost. The newest version of GPT-3 is the most expensive model, but also the most powerful one from OpenAI. Since the release of GPT-3, however, the OpenAI transformers are closed-source and not available as stand-alone software. OpenAI provides a restricted access for the most sophisticated versions via an API and only for specific use cases. Because of these restrictions, the researcher group EleutherAI provides GPT-NeoX as free available alternative [3]. GPT-NeoX has less parameters than GPT-3, but it has been trained on more data. The English training data set is also called "*The Pile*" [7] and has been applied for training state-of-the-art models. Up to now, GPT-NeoX has only been trained for completing given text phrases written in English. There are also German GPT-Models [8, 9]. The German GPT-2 Version has been released in 2020, but it has been trained with a rather small set of training data [8]. The "German-GPT2-Larger-Model" was released 2021 and has been trained on more data, but it is based on the GPT-2 transformer architecture too. Since the beginning of 2022, there is also an European GPT-Model, called OpenGPT-X [13], which has as much parameters as the GPT-3-Model. Its total cost is about 15 million dollars. OpenGPT-X is still not open-source. The access is restricted, similar to GPT-3.

Table 1.1 State-of-the-art GPT-models

	Model	Data	Costs (US$)	Parameters	Release	Availability
OpenAI	GPT-2 [14]	40 GB	$>50 * 10^3$	$1{,}542 * 10^6$	2019	Open-source
OpenAI	GPT-3 [4]	570 GB	$>4 * 10^6$	$175 * 10^9$	2020	Closed-source online-API
DBMZ [8]	GPT-2 (germ.)	16 GB	n.a.	n.a.	2020	Open-source
Stefan-it [9]	GPT-2(germ.)	90 GB	n.a.	n.a.	2021	Open-source
EleutherAI	GPT-NeoX [3]	800 GB	n.a.	$>20 * 10^9$	2022	Open-source
German institutions	OpenGPT-X [13]	n.a.	$15 * 10^6$	$>175 * 10^9$	2022	Closed-source

1.3 Chat Generation Framework

The aim of our chat generation framework is to generate communication flows between two autonomously acting bots in different languages and to transform them into required data type formats. The communication topic can be specified by the user. The communication flow can be affected by various parameters. In addition, domain-specific content can be included. The generation process consists of two phases: the actual data generation and the conversion into the required data formats (see Fig. 1.1).

1.3.1 Test Data Generation

For test data generation, we have adopted the open-source Python-based human–machine chatbot from the *transfer-learning-conv-ai* project [22]. It is based on a GPT-model and supports already topic-oriented chats. We have chosen the GPT-2 model because it is open-source as our framework should be for verifiability reasons. This also enables an extension with other open-source tools. Another important selection aspect has been the trainability of the model for chat generation.

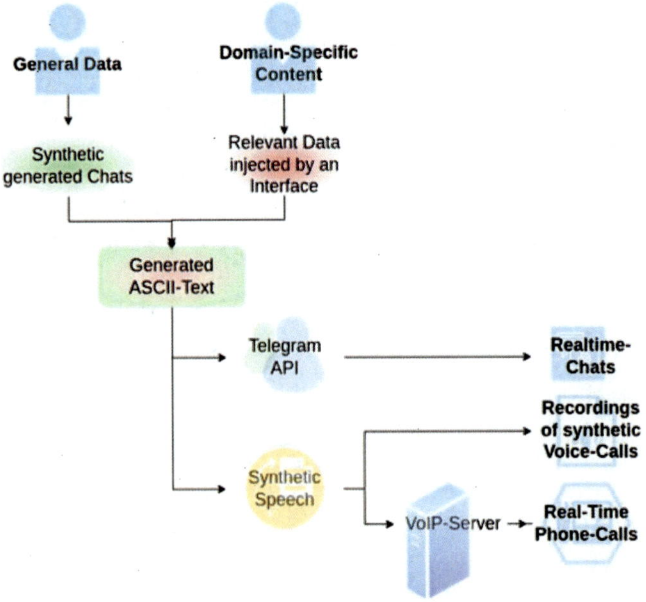

Fig. 1.1 Test data generation workflow

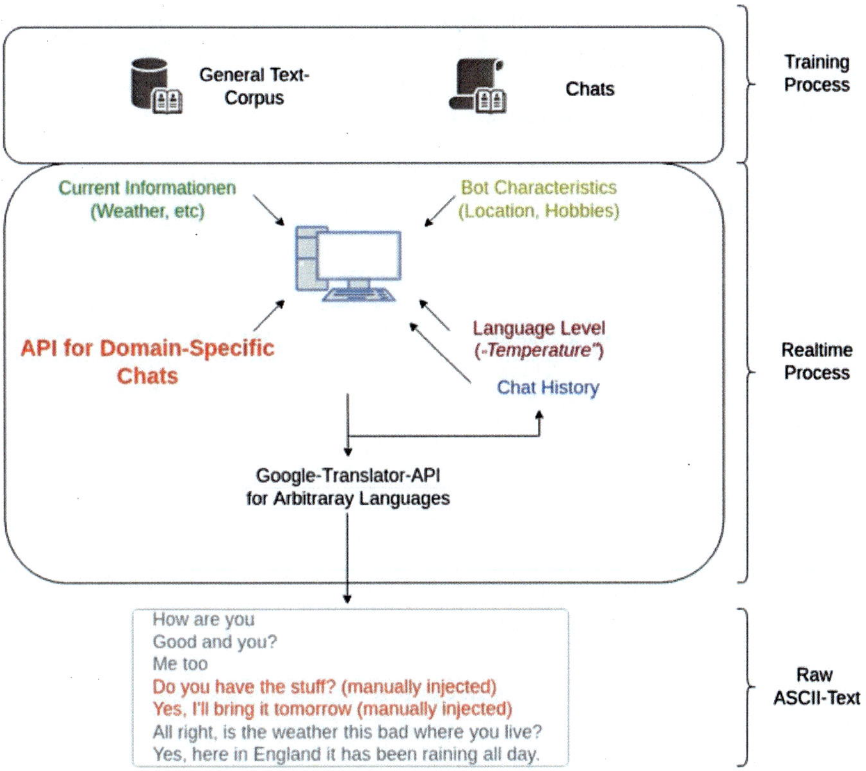

Fig. 1.2 Chat generation

Chat Generation

The training of a transformer for chat generation consists of two steps (see Fig. 1.2). In the first step, it is trained on a data set with general texts like documents and digital books for understanding the target language. This step has already been adapted for all transformers as mentioned in Chap. 1.2.2. In the second step, the transformer is fine-tuned for generating chat texts. For this, a big amount of language-specific chat training data is required. There is only a small number of such data sets, mainly in English. For the generation of chats in other languages, this is a problem. We discuss this further below. Therefore, we applied the GPT-2 model for English. The fine-tuning was performed with the already mentioned *Persona-Chat* data set published in 2018 [23]. It consists of 162,064 English utterances in a characterizing person-to-person context. If a chat is to be generated, the topic of the conversation between the bots must first be determined (see Fig. 1.2). This is done with the bot characteristics (yellow) that defines the topic, e.g., ordering a pizza, but it also provides additional information, such as the names of the bots, age, or location. In addition, other information (green) may be added for more realistic chats, e.g., current weather information from a free API. To generate responses, the model needs the current chat message which it should answer. Furthermore, the history of the last chat message exchanges is taken into account to better adapt the context. The length of the chat history can be parameterized.

Adjusting the Language Level

The quality of the generated chats can be influenced via various parameters. The most relevant parameter for the conversion style is called "*temperature*" of the model (see Fig. 1.2). It determines the probability (q_i) of each token in the model to be the next word in the chat. This probability is defined as follows:

$$q_i = \frac{e^{\left(\frac{z_i}{T}\right)}}{\sum_j e^{\left(\frac{z_j}{T}\right)}}. \tag{1.1}$$

q_i	new property for choosing token i
z_i	output value of the NLP model for token i
T	temperature
$\sum_j$	sum over all available tokens

The temperature controls how much the probability of the next word, in this context defined as token, is oriented to the direct output of the NLP model (z_i). A temperature value of 1.0 does not affect the selection process. With a higher temperature, the probabilities of the different tokens become more similar; accordingly, a lower temperature causes a harder cut between the probabilities of the tokens. This control can be used for generating probability-oriented tokens for a formal language, e.g., "Good Morning" with 0.7. A low value causes a more formal speech, whereas a higher temperature, e.g., 1.2, leads to a more informal speech, e.g., "Hey Bro." As a consequence, a low temperature prefers the word combinations that are usually included in the training data set. In the case of GPT-2, this is the intersection of all training data, including the general text corpus with the Wikipedia articles. The intersection of these data sets consists of syntactically correct standard phrases that can appear in documents or in formal chat phrases. These combinations have the highest probability in comparison to rare specialized informal situational chats that appear only in specialized training data sets, e.g., chats in dialects to special topics. A high temperature instead allows these seldom word combinations to appear and generates text phrases that are rather contextual.

Injection of Domain-Specific Chats

As already mentioned above, the phrases or chats, respectively, generated by the GPT-model for a given topic are nevertheless general and in part accidental. Forensic data sets also require dialogues that relate to certain topics or cases, e.g., crime-related conversation. To introduce such domain-specific dialogues into the chat generation, we have implemented an API for injecting clear defined messages (see Fig. 1.2). Thus, irrelevant small talk, for instance, can be combined with hate speech, fake news, conspiracy talks, etc. An example of manually injected hate chats (red) is given in the chat example contained in Table 1.5. These injected data can then be used to test extraction functionalities, e.g., relevant comments, in forensic analysis tools.

Translation to Other Languages

Our goal is to provide a multi-language chat generation framework to generate besides English also chats in our native language German and in other languages. Since there are scarcely chat training data and pretrained transformers in other languages, we were compelled, as explained above, to use English-trained ones that consequently generate texts in English. An automated translation of the training data would decrease the quality of the applied model. Therefore, we did not use this option and applied another way. We decided to use a real-time translation of the generated English chat (see Fig. 1.2). This option can be applied to any language the translation model supports. Thus, our framework can be updated by any GPT-model that has been trained for generating chat messages. The precondition for using larger transformer models is a higher amount of needed chat training data caused by a higher number of parameters to be set. This challenge can only be tackled with English because in this language space there are more active research groups and a greater demand for new training data sets than in other languages. We have used the free API of the Google Translator. It allows a fast online translation without using an account and can be included in Python-based implementations. The translator supports 108 languages. Thus, our framework supports currently each of the 108 languages. Because of the simple short structure of most chat messages, a translation does not cause a larger quality loss. The translation is required in two phases of the model: 1. For the API, for manually injected domain-specific phrases to provide the model with the English version of the phrase. 2. After the next word was predicted by the transformer in the trained (English) language.

1.3.2 Data Formatting

Instant Messaging APIs

Chat generation produces raw ASCII chats. For forensic use cases, these files have to be converted into the required data format and enriched with metadata, e.g., time stamps. Data formatting is therefore the second important step in chat generation. For this, a possibly wide range of formats should be supported. Hate speech, for example, is often sent over the application *Telegram*. To generate equivalent test data, we implemented the Telegram API in the chatbot [1] (see Fig. 1.1). It enables controlling the Telegram app by automated Python scripts. Thus, messages can be sent, received, and analyzed without human interaction. In combination with the synthesized text, it allows us to automatically send chatbot messages between two smartphones identically to real-world Telegram conversation.

Synthesizing Speech

Besides written chats, it is often necessary to also support use cases that consider tests in speech analysis or speech-to-text functionalities. We provide this option too and transform the chats into audio files that allow us to provide artificial telephone calls (see Fig. 1.1). For this, we used open-source audio generating tools. We adopted the research project "Real-Time-Voice-Cloning" [10, 15]. It allows us to clone any given English voice character in real time. The project uses different neural network architectures of previous projects, such as vocoder [11], synthesizer [21], and encoder [20] trained with English. For German, we used a forked project that has retrained the models with German speakers [16]. Because of the lower number of German training data, the German version is not able to imitate any speaker in real time. Our tests have shown though that it is possible to use fine-tuning of the synthesizer for adjusting it to a given speakers voice. We used *wav* files from the M-Ailabs data set [18]. The result is a complete set of *wav* files that contain the spoken representations of the generated texts voiced by predetermined speakers. These *wav* files can be exported and used for forensic tests. For extending the *wav* files with meta data, we have deployed *Asterisk* [17] as VoIP server. The server is configured for automatically responding to voice chats, oriented at the "Lenny-Project" [6]. These accounts can be combined for fully automated voice chats. The data can be extracted by network traffic detection tools, e.g., *Wireshark*. It supports all relevant voice-over-IP (VoIP) protocols such as the real-time transport protocol (RTP), the real-time control protocol (RTCP), and the session initiation protocol (SIP). As a consequence, the voice traffic is identical to real calls and can be used for further analyses or extracting procedures.

1.4 Evaluation

We performed various evaluations to estimate the conversational quality of the generated chats and to test the compatibility of the data formats. Note that we only consider the generated chats here. Speech synthesis is currently limited exclusively to functional feasibility.

1.4.1 Qualitative Analysis of the Generated Chats

The goal of this evaluation was to estimate the conversational quality of the generated chats. We considered three chat scenarios for which we used essential features of our framework. The chats were generated in English. They can be translated into other languages as described above. The chatbots were provided with only limited knowledge about themselves in the bot characteristics (see Table 1.2). This knowledge may contain permanent information about the agent, e.g., name, age, or temporary information. Our tests showed that our framework can generate up to 16 consistent realistic chat messages. The higher the number of chat messages of a conversion, the higher becomes the risk of inconsistent contents, e.g., contradictory statements to the bot names or locations. In the generated messages appear syntactic errors such as the wrong spelling of the word *"but"* in Table 1.4. Furthermore, most of the words were generated in lowercase. These effects are caused by the training data and simulate a realistic structure of chat messages.

The first scenario represents a normal chat without any domain-specific information included in the conversation. The resulting chat, generated with the standard language-level temperature of 0.7, is presented in Table 1.3.

As it can be seen, the conversation consists of grammatical correct phrases about weather, hobbies, age, and the location of the chat's participants. The statements pick up the pre-defined topics provided in the bot's characteristics (cf. Table 1.2). The generated chats do not contain repetitive or contradictory statements. Table 1.1 also shows that the chatbots can produce synthetic questions and matching answers such as the question about their hobbies. Thus, the conversation becomes realistic and consistent. The second scenario changes the language level to a more informal, conversational style. It simulates a chat with a more chat like language and more psychological characteristics, e.g., the mood of the partners or the type of

Table 1.2 Bot characteristics used for the generated chats

Bot	Description used for a specific content
Bot A	My name is Max
	I am old
	It's a bit cold
Bot B	My name is Andy
	I am in China !
	I'ts still a bit cold

Table 1.3 Synthetically generated chat messages

Bot	Message
Bot A	I'm really old.
Bot B	I'm 36.
Bot A	I've never had cold weather.
Bot B	I've been to china a few times.
Bot A	What do you do for a living?
Bot B	I work in a clothing store.
Bot A	Do you have any hobbies?
Bot B	I like to hang out with my friends and dogs.
Bot A	I like to hike, snowboarding, fishing, fishing.
Bot B	I love fishing too.

Table 1.4 Synthetically generated chat messages with temperature 1.2

Bot	Message
Bot A	Hi, hello. here tell me about you?
Bot B	Hello! here tell me about yourself!
Bot A	I'm big on an ice cold facebook live feed!
Bot B	Cool! i'm in china still... you?
Bot A	Funny stuff! who played guitar until you break ice cold?
Bot B	Yeah played me and ugh yeah
Bot A	Nice! wish i went to china! we definitely weren't other international lines anyway.
Bot B	Yes bet that would be strange. i'm a professional cobbler

conversation. The generated chat is presented in Table 1.4. It was generated with an increased language-level temperature of 1.2. The chat contains more exclamations and informal expressions, such as "yeah," "funny stuff," and "ugh." Hence, the conversation has more emotional phrases and is looser. It also has ambiguous sentences, such as *"yeah played me and ugh yeah."* These phrases are typical for real-life chat conversations that often contain short emotional comments without pursuing formal syntactic rules.

The third scenario is a regular chat in which three chat messages with domain-specific, case-relevant information are added (see Table 1.5). The manually injected phrases are marked in red. The injected statements are saved in the chat history to use the manually injected phrases for future automated responses. The scenario shows that manually adding chat messages to the chat history does not negatively impact the generation of realistic chats. The subsequent synthetic chats have other content but are still consistent with previous synthetic and manual responses. A complete independence between the manually injected and synthetic chats is possible by omitting the manually generated chat in the chat history. However, this would increase the risk of contradictions between the manually and synthetically generated phrases.

Table 1.5 Synthetically generated chat messages (black) injected with relevant data (red)

Bot	Message
Bot A	What do you do for a living?
Bot B	I'm a teacher. I teach English in China)
Bot A	Do you think Santa Clause is a big conspiracy?
Bot B	Yes of course, these idiots just want to rule the world!
Bot A	Exactly!
Bot B	I am not very good at English.
Bot B	I'm not good at it either

Table 1.6 Use cases and synthetically generated data types, applicability marked by *"X"*

	Test data for tools for		
Data type	Semantic analysis	Speech processing	Data extraction
ASCII-text	**X**		
Network traffic			**X**
Telegram messages	**X**		**X**
Wav files		**X**	
VoIP data		**X**	**X**

1.4.2 Applicability of the Generated Data Formats

The framework in its current state supports different data formats (see Table 1.6) that focus on different application areas. The raw ASCII text generated at the beginning can be used for text analyzing tools without complex extracting algorithms. The framework supports the transformation into the other data formats of Table 1.6. The Telegram format used in our experiments with the Telegram app is more realistic but harder to analyze because of the encryption and encoding used in the app. The contained metadata, such as time stamps or telephone numbers, are saved on the smartphones and can be extracted as hard drive image or from the Telegram database. *Wav* files contain raw audio data and can be applied as test data for speaker detection ad speech-to-text tasks. In addition, *wav* files can be included in more complex VoIP applications. The metadata contained in this traffic can be used for testing network analyzing tools.

1.5 Conclusion

In this chapter, we have presented a framework for automatically generating chats on a given topic for forensic purposes. It is based on chatbots. Unlike human-to-bot systems, it supports a bot-to-bot communication that allows to simulate real-life chat and telephone communication. The generated chats are provided in different data and audio formats. They can be translated to any language if it is supported by the translation API, and there is a training data set for a text-to-speech transformation in the case of audio chats. The framework allows us to introduce specific statements into the communication to produce case-relevant forensic data. For this purpose, the framework provides an interface for injecting domain-specific phrases. The framework applies NLP methods and open-source tools. They are configurable and ensure that every generation step is explainable and can be expanded by novel features. For the audio chats, different speakers and arbitrary chosen text contents can be used. For generating huge amounts of test data, the synthetic chat generation can be automatized to generate message samples of a given length, language, and topic. Based on these samples, semantic analysis methods and forensic tools can be evaluated. For testing extracting tools, domain-specific messages can be injected, sent over the network, and compared with the messages extracted by the tool then. Generated speech files can be used for testing speech-to-text algorithms by giving them the generated audio chats and comparing the output with the original chat messages. Further functionalities such as (meta)data extraction and format interpretation can be evaluated by using the framework's ability of generating Telegram messages. The framework is completely modular. This allows us to add further formatting tools, i.e., for audio data and social network data in our further research. In a next step, we plan to enable conversations with more than two partners. This requires a new fine-tuning training and a more complex transformer model. After successfully evaluating the proof-of-concept with GPT-2, we want to extend our framework with the online API of GPT-3, the Open-GPT-X project, or an implementation of the GPT-Neo-X project to further increase the quality of the generated texts. This should allow us to generate messages over a long time under consistency considerations and without statement duplication.

References

1. A. Akhmetov, Python-telegram, https://github.com/alexander-akhmetov/python-telegram. Accessed 24 Oct 2022
2. T.A. Almeida, J.M.G. Hidalgo, A. Yamakami, Contributions to the study of SMS spam filtering: new collection and results, in *DocEng '11* (2011)
3. S. Black, L. Gao, P. Wang, C. Leahy, S. Biderman, GPT-Neo: Large scale autoregressive language modeling with mesh-tensorflow (2021). https://doi.org/10.5281/zenodo.5297715
4. T. Brown, B. Mann, N. Ryder, M. Subbiah, J.D. Kaplan, P. Dhariwal, A. Neelakantan, P. Shyam, G. Sastry, A. Askell, et al., Language models are few-shot learners, in *Advances in Neural Information Processing Systems*, ed. by H. Larochelle, M. Ranzato, R. Hadsell, M.F. Balcan, H. Lin, vol. 33 (Curran Associates, Inc., Red Hook, 2020), pp. 1877–1901. https://proceedings.neurips.cc/paper/2020/file/1457c0d6bfcb4967418bfb8ac142f64a-Paper.pdf
5. W.W. Cohen, other: Enron email dataset, https://www.cs.cmu.edu/~enron/. Accessed 24 Oct 2022
6. S. Ewing, Introducing Lenny, https://shaun.net/notes/introducing-lenny/. Accessed 24 Oct 2022
7. L. Gao, S. Biderman, S. Black, L. Golding, T. Hoppe, C. Foster, J. Phang, H. He, A. Thite, N. Nabeshima, et al., The Pile: an 800GB dataset of diverse text for language modeling. Preprint. arXiv:2101.00027 (2020)
8. German GPT-2 model, https://huggingface.co/dbmdz/german-gpt2. Accessed 24 Oct 2022

9. German GPT-2 model large, https://huggingface.co/stefan-it/german-gpt2-larger/. Accessed 24 Oct 2022
10. Y. Jia, Y. Zhang, R. Weiss, Q. Wang, J. Shen, F. Ren, Z. Chen, P. Nguyen, R. Pang, I. Lopez Moreno, Y. Wu, Transfer learning from speaker verification to multispeaker text-to-speech synthesis, in Advances in *Neural Information Processing Systems*, ed. by S. Bengio, H. Wallach, H. Larochelle, K. Grauman, N. Cesa-Bianchi, R. Garnett, vol. 31 (Curran Associates, Inc., Red Hook, 2018). https://proceedings.neurips.cc/paper/2018/file/6832a7b24bc06775d02b7406880b93fc-Paper.pdf
11. N. Kalchbrenner, E. Elsen, K. Simonyan, S. Noury, N. Casagrande, E. Lockhart, F. Stimberg, A. van den Oord, S. Dieleman, K. Kavukcuoglu, Efficient neural audio synthesis, in *Proceedings of the 35th International Conference on Machine Learning. Proceedings of Machine Learning Research*, ed. by J. Dy, A. Krause, vol. 80 (PMLR, 2018), pp. 2410–2419. https://proceedings.mlr.press/v80/kalchbrenner18a.html
12. A. Köhn, F. Stegen, T. Baumann, Mining the spoken Wikipedia for speech data and beyond, in *Proceedings of the Tenth International Conference on Language Resources and Evaluation (LREC 2016)*, ed. by N.C.C. Chair, K. Choukri, T. Declerck, M. Grobelnik, B. Maegaard, J. Mariani, A. Moreno, J. Odijk, S. Piperidis (European Language Resources Association (ELRA), Paris, 2016)
13. L. Lehmhaus, OpenGPT-X, https://www.aleph-alpha.com/the-next-press-release
14. A. Radford, J. Wu, R. Child, D. Luan, D. Amodei, I. Sutskever, Language models are unsupervised multitask learners. OpenAI blog **1**(8), 9 (2019)
15. Real-time-voice-cloning, https://github.com/CorentinJ/Real-Time-Voice-Cloning. Accessed 24 Oct 2022
16. Real-time-voice-cloning-German, https://github.com/padmalcom/Real-Time-Voice-Cloning-German, Accessed 24 Oct 2022
17. Sangoma Technologies, Asterisk, https://www.asterisk.org/. Accessed 24 Oct 2022
18. I. Solak, M-Ailabs, https://www.caito.de/2019/01/03/the-m-ailabs-speech-dataset/. Accessed 24 Oct 2022
19. A. Vaswani, N. Shazeer, N. Parmar, J. Uszkoreit, L. Jones, A.N. Gomez, L.u. Kaiser, I. Polosukhin, Attention is all you need, in *Advances in Neural Information Processing Systems*, ed. by I. Guyon, U.V. Luxburg, S. Bengio, H. Wallach, R. Fergus, S. Vishwanathan, R. Garnett, vol. 30 (Curran Associates, Inc., Red Hook, 2017), https://proceedings.neurips.cc/paper/2017/file/3f5ee243547dee91fbd053c1c4a845aa-Paper.pdf
20. L. Wan, Q. Wang, A. Papir, I.L. Moreno, Generalized end-to-end loss for speaker verification. ICASSP (2018), https://arxiv.org/abs/1710.10467
21. Y. Wang, R.J. Skerry-Ryan, D. Stanton, Y. Wu, R.J. Weiss, N. Jaitly, Z. Yang, Y. Xiao, Z. Chen, S. Bengio, Q.V. Le, Y. Agiomyrgiannakis, R.A.J. Clark, R.A. Saurous, Tacotron: towards end-to-end speech synthesis, in *INTERSPEECH* (2017)
22. T. Wolf, V. Sanh, J. Chaumond, C. Delangue, TransferTransfo: a transfer learning approach for neural network based conversational agents. abs/1901.08149 (2019), http://arxiv.org/abs/1901.08149
23. S. Zhang, E. Dinan, J. Urbanek, A. Szlam, D. Kiela, J. Weston, Personalizing dialogue agents: I have a dog, do you have pets too? in *Proceedings of the 56th Annual Meeting of the Association for Computational Linguistics* (Volume 1: Long Papers), Melbourne (2018), pp. 2204–2213. https://doi.org/10.18653/v1/P18-1205, https://www.aclweb.org/anthology/P18-1205

Chapter 2
Understanding Responses to Embarrassing Questions in Chatbot-Facilitated Medical Interview Conversations Using Deep Language Models

Ching-Hua Chuan, Wan-Hsiu S. Tsai, Di Lun, and Nicholas Carcioppolo

2.1 Introduction

Healthcare has been a popular domain for CAs, from the earliest Eliza [1] that simulates a psychotherapist to the recent Clara from the US Centers for Disease Control and Prevention that answers questions about COVID-19 symptoms. Apart from being interactive and accessible, such automated conversational interfaces have a unique potential for facilitating health-related conversations, especially on topics that may be embarrassing for people to discuss with their doctors, due to concerns of personal privacy or social stigma. As suggested by Weisband and Kiesler [2], "people would tell an impartial machine personal or embarrassing things about themselves, without fear of negative evaluation" (page 3). In this case, chatbots can be an effective platform to encourage self-disclosure of important personal information and honest description of experiences or concerns regarding a health issue.

It is important to note that a chatbot is more than a simple replacement for medical questionnaires. Powered by artificial intelligence and machine learning, chatbots can understand the conversation content to respond in an appropriate manner. This capability is particularly crucial for medical interviews because this

C.-H. Chuan (✉)
Department of Interactive Media, University of Miami, Coral Gables, FL, USA
e-mail: c.chuan@miami.edu

W.-H. S. Tsai
Department of Strategic Communication, University of Miami, Coral Gables, FL, USA
e-mail: wanhsiu@miami.edu

D. Lun · N. Carcioppolo
Department of Communication Studies, University of Miami, Coral Gables, FL, USA
e-mail: dxl744@miami.edu; n.carcioppolo@miami.edu

M. D. Hina et al. (eds.), *Future Research Directions in Computational Intelligence*,
EAI/Springer Innovations in Communication and Computing,
https://doi.org/10.1007/978-3-031-34459-6_2

allows the chatbot to actively seek clarification or elaboration with the user and provide empathy to ease emotional stress, thereby improving user experience [3].

In this chapter, we examine two state-of-the-art deep language models on classifying the level of detail and the presence of self-disclosure in conversations with a medical interview chatbot about colorectal health. Prior studies have explored machine learning techniques on detecting self-disclosure [4]. However, these studies focus on social media posts or online support group forums. No research to date has studied self-disclosure in the medical interview setting that contains embarrassing questions. To collect conversational data, we created a chatbot that is embedded on a colon health information website to conduct medical interviews with study participants. We first analyzed the data by exploring the conversation in terms of word count distribution and word cloud representations for each class. We then fine-tuned the BERT model and used the newly released GPT-3 classification API to classify the conversation based on different levels of detail and whether participants' input contained self-disclosed information. The results were examined via accuracy against results of content analysis conducted by two human coders as the ground truth. At last, we shared some promising findings and surprising results in the discussion session.

2.2 Colon Health Chatbot

2.2.1 Chatbot Design

Recognizing that embarrassment is a major barrier to colon cancer prevention [5], this study designed a medical interview chatbot for colorectal health. To simulate a medical interview, the questions asked by the chatbot contained several yes/no and multiple-choice questions regarding the participant's personal experience and condition related to colon health. If the participant answered yes, the chatbot would then ask the participant to elaborate. Some questions asked by the chatbot were designed to induce feelings of embarrassment, such as asking participants to describe the color and size of their recent stool, experiences with diarrhea, constipation, anal sex, and colonoscopy. The chatbot was programmed in JavaScript such that the participant can interact with the chatbot on a web interface by either clicking a button or typing their answers. A snippet of conversation exchanges between a participant and the chatbot is presented in Fig. 2.1.

2.2.2 Data Collection and Content Analysis

A total of 552 participants who were 35 years or older were recruited via Qualtrics' online panel and completed the medical interview. Participants were instructed to

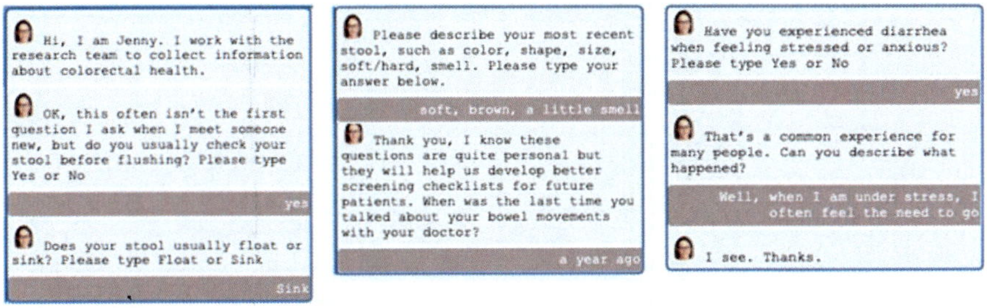

Fig. 2.1 A snippet of conversation exchange between the chatbot (text in black) and a participant (text in white)

briefly browse the website on colorectal cancer and then to chat with an online medical assistant to answer questions about their colon health. The participants spent between 3 and 40 minutes on the conversation, with the average time around 10 minutes.

Two human coders who were not the authors were trained and coded the medical interview conversations. For the level of detail, a message was coded as *not detailed* if it only provided vague, general answers (e.g., "as usual"), as *somewhat detailed* if some details were provided (e.g., "stool is brown, about 4-inch long, smell"), and as *very detailed* if it contained many details ("it was medium brown, not especially dark or light. Maybe an inch in diameter and 4 inches long. Not hard and not soft and runny. Little smell that I remember."). Regarding self-disclosure, a message would be coded as containing self-disclosure when it included health information not directly related to colon health (e.g., "I was born with poly cystic liver disease") or personal information (e.g., "I am retired"). Three rounds of coding were performed and the inter-coder reliability for the final coding reached above 0.7 using Cohen's Kappa calculation (0.715 for details and 1 for self-disclosure).

2.3 Exploring Medical Interview Conversations

2.3.1 Level of Detail

A total of 3287 responses to open-ended elaboration questions were collected, with 826 coded as very detailed, 1721 somewhat detailed, and 740 not detailed. The distribution of word count in the response for each detail category is illustrated in Fig. 2.2, using a non-parametric kernel density estimate plot. Longer responses tend to contain more details than shorter ones. However, word count is not a determinant factor for the level of detail in the conversation because the distributions of the three categories overlap as shown in the figure.

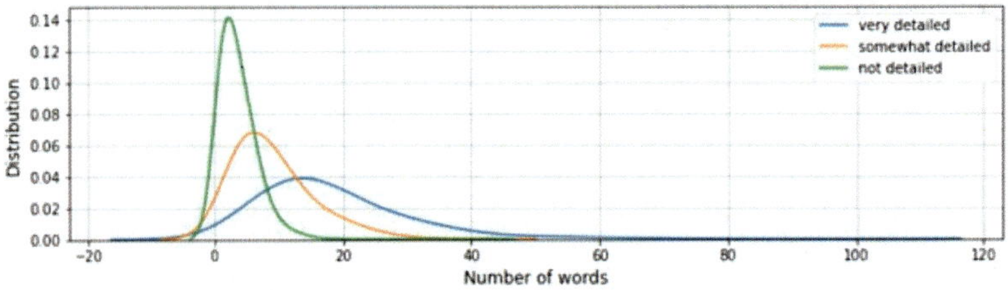

Fig. 2.2 Kernel density plot on the number of words in responses for different levels of detail

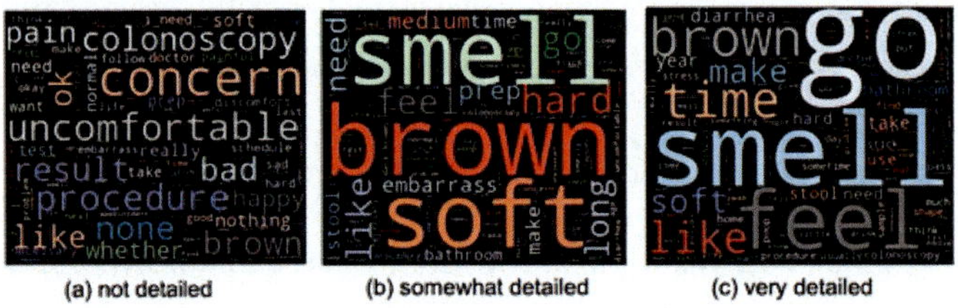

Fig. 2.3 Word cloud representations for responses in three levels of detail

To examine the content for the three levels of detail, word cloud representations were created in Fig. 2.3 for (a) not detailed, (b) somewhat detailed, and (c) very detailed.

The word cloud was generated after pre-processing steps including tokenization, removing stopwords, identifying noun phrases, and lemmatization. The total number of unique words in each category are (a) 2660, (b) 14,388, and (c) 13,616. The difference between not detailed and somewhat/very detailed is visible: (b) and (c) contain words such as smell, brown, and soft that provide specific details about stool, while (a) includes more words related to colonoscopy concerns. In contrast, the difference between (b) and (c) is less obvious.

2.3.2 *Self-Disclosure*

All participants (552) responded in the chat for at least one optional elaboration question. Out of all participants' responses (total 3287), only 101 were coded as containing self-disclosure. Figure 2.4 shows the distribution of word count for responses with and without self-disclosures. As shown in the figure, whether or not a conversation contains self-disclosure cannot simply be determined by the length of text in the chat.

Figure 2.5 shows the word clouds for responses (a) with and (b) without self-disclosures. In responses with self-disclosures as shown in (a), words related to health conditions (e.g., colon cancer and polyps) and words used to describe their personal history (e.g., year, time, since, ago) can be observed. In contrast, most

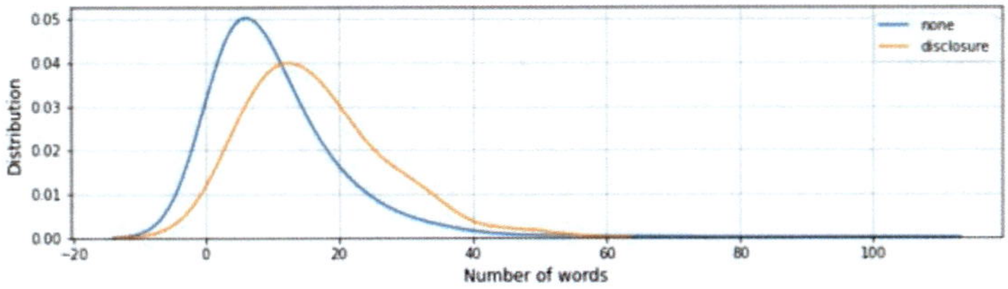

Fig. 2.4 Kernel density plot on number of words in responses with and without self-disclosure

Fig. 2.5 Word cloud representations for responses (**a**) with and (**b**) without self-disclosure

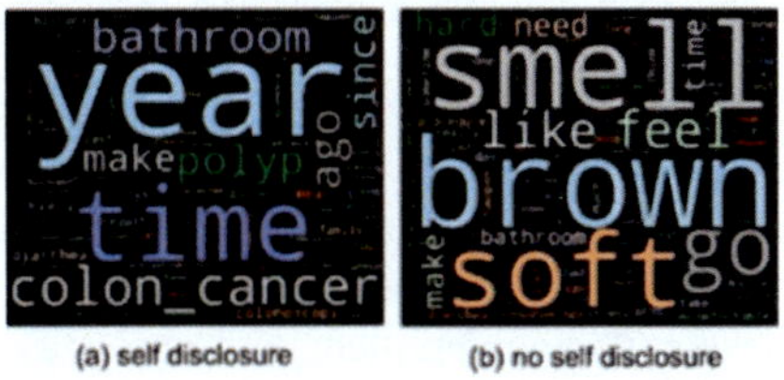

words in responses without self-disclosure as shown in (b) are generally related to stool descriptions.

2.4 Classification and Results

2.4.1 Language Models: BERT and GPT-3

The most prominent breakthrough in natural language processing (NLP) in recent years is transformer-based models. The transformer architecture with the attention technique [6] solves the key problems of other deep learning models in NLP because of its capability to model long-term context and provide parallelization for scalability. In particular, the Bidirectional Encoder Representations from Transformers (BERT) proposed by Devlin et al. [7] demonstrated how a pre-trained transformer model can be effectively fine-tuned with just one last output layer for a variety of machine learning tasks. In this study, we used the base BERT model from the Hugging face library and fine-tuned the last layer in PyTorch for the level of detail and presence of self-disclosure classification.

Another transformer-based model that has captured a lot of attention is the Generative Pre-trained Transformer 3 (GPT-3) from OpenAI, the largest pre-trained language model to date. However, the pre-trained model is not directly available to the public; instead, OpenAI provides a "toolset" of GPT-3 APIs for various tasks including text generation, Q&A, language translation, and classification, which are used in this study. Instead of retraining the model, the GPT-3 classification API performs meta-learning or in-context learning, which determines the output based on a few given examples of text and their labels. To use the API, we first prepared the training set as a jsonl file and uploaded it to the server. The classification

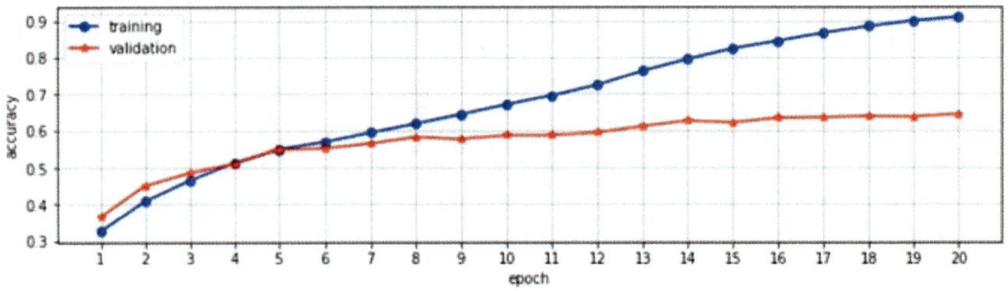

Fig. 2.6 Cross-validation accuracy of the fine-tuned BERT model for detail classification

API endpoint is a combination of Search and Completion: first, a keyword Search is performed to find the similar examples from the training set. The number of maximum examples for the search can be defined in the API call. Once the similar examples are identified, the Completion endpoint reranks the examples via semantic search to determine the label of a query text.

The following two sections describe how BERT and GPT-3 models were tested. The findings for classifying medical interview conversations in terms of level of detail and presence of self-disclosure are also discussed.

2.4.2 Results: Level of Detail

To prepare the data for classification of the level of detail, all responses were separated into training and validation sets using a ten-fold cross validation procedure with stratified sampling. The same 10 sets of training and validation data were used for BERT and GPT-3 classifiers. Figure 2.6 shows the training and validation accuracy of the fine-tuned BERT model (Adam optimizer on cross entropy loss, learning rate = 1e-6, dropout rate = 0.5, batch size = 32). As shown in the figure, the validation accuracy stays around 0.6 after 8 epochs while training accuracy keeps increasing. To understand how well the model works in each of the three detail levels, a confusion matrix (normalized by the true label) is presented in Fig. 2.7a. In general, the model works the best for identifying "somewhat detailed" responses (0.73), but the model also tends to misclassify responses from the other two categories as "somewhat detailed." In addition, the model seldom misclassified cases between "not detailed" and "very detailed" (0.047 and 0.054), which is a promising potential of using this model as a classifier.

For the GPT-3 classifier, the 10 training sets were uploaded to the GPT-3 server and used for the corresponding validation set, in which each example was included in a query and sent to the GPT-3 classification API to obtain the label. The training-and-validation process was conducted twice, one with the search maximum set to 30 and the other to 100 examples. The confusion matrix for the GPT-3 classifier is shown in Fig. 2.7b and c. The results show that the GPT-3 classifier overwhelmingly

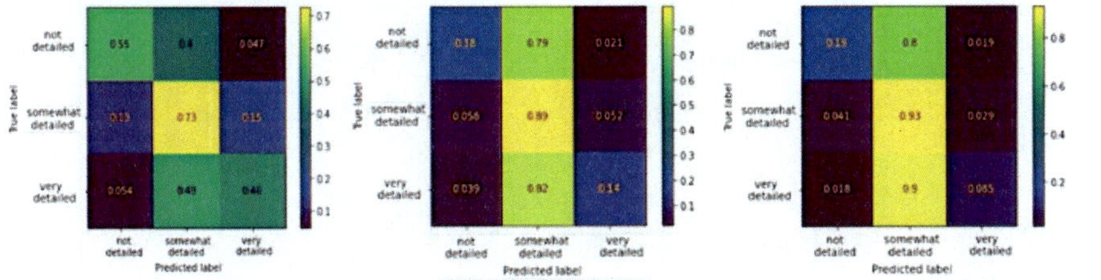

(a) BERT Transformer **(b) GPT-3 Classifier (30 examples)** **(c) GPT-3 Classifier (100 examples)**

Fig. 2.7 Confusion matrix for detail classification using (**a**) BERT, (**b**) GPT-3 classifier with 30 examples, and (**c**) GPT-3 classifier with 100 examples

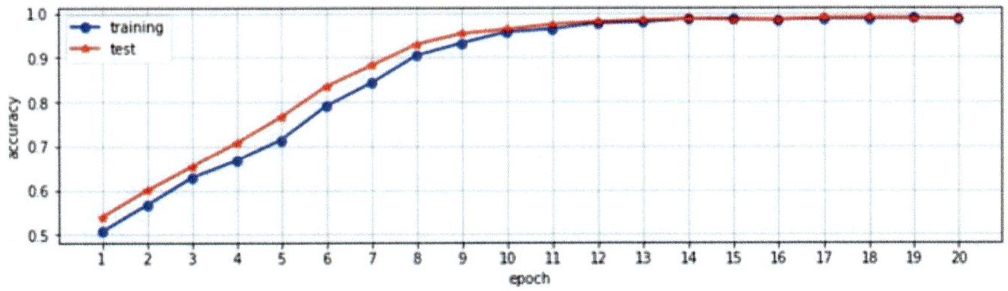

Fig. 2.8 Training and test accuracy of the fine-tuned BERT model for self-disclosure classification

labeled the examples as "somewhat detailed" across the three levels, and increasing the maximum example for search did not improve the accuracy.

2.4.3 *Results: Self-Disclosure*

Since only 101 out of 3287 responses were labeled with self-disclosure present, stratified sampling will result in very skewed distributions. Therefore, we included all examples with self-disclosure, and randomly selected an equal number of examples without self-disclosure to form a dataset. The dataset was then randomly split into a training set (80%) and test set (20%). We repeated the process 10 times to create 10 different training and test sets for the classification.

Figure 2.8 shows the training and test accuracy of the fine-tuned BERT model for self-disclosure classification. As shown in the figure, the fine-tuned BERT model works extremely well, as the accuracy approached to almost 1 after 14 epochs.

The confusion matrix for GPT-3 classifier on self-disclosure classification is shown in Fig. 2.9. Compared with the fine-tuned BERT model, the GPT-3 classification API performed poorly by reporting a lot of false positives (0.46). Increasing the number of search examples did not improve the overall accuracy.

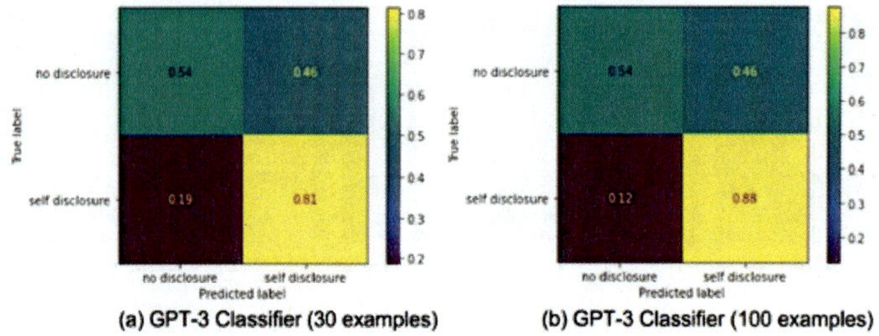

Fig. 2.9 Confusion matrix for GPT-3 classifier with (**a**) 30 and (**b**) 100 examples on self-disclosure classification

2.5 Conclusion and Discussion

This exploratory study utilized a multidisciplinary approach to examine the effectiveness of deep language models for understanding conversations in chatbot-facilitated medical interviews involving embarrassing topics. A web-based chatbot was created to conduct medical interviews interactively with users on colorectal health. The chatbot successfully delivered the interview by maintaining the conversation with participants on the average of 10 minutes. Responses from more than five hundred participants were recorded and manually coded for the level of detail and the presence of self-disclosed information. Deep language models of BERT and GPT-3 were tested to classify the response against the human labeled ground truth.

For both classification tasks, the fine-tuned BERT model performed better than GPT-3 classification API. The poor performance of GPT-3 classifier may be due to the keyword search in the first step, which reduced the entire training set to a smaller set of examples that contain the same keywords without considering contextual information. In addition to the GPT-3 classification API, OpenAI also offers other options, such as word embedding and fine-tuning, that may perform better for this task.

Compared with the work by Valizadeh et al. [4], which tested a lightweight BERT model on medical self-disclosure classification of 6639 posts from online social platforms and reported the accuracy of 81%, conversations in medical interviews are relatively shorter in length and more focused on a specific health issue. The perfect performance of the BERT language model on self-disclosure classification in this study highlights the promising advantages of BERT in analyzing medical interview conversations.

Chatbots are increasingly adopted for healthcare purposes, especially for embarrassing and potentially stigmatizing medical conversations such as mental, sexual, and colon health [8] when identifying and understanding self-disclosures are imperative. The study findings highlight that incorporating language models such as BERT when creating chatbots for sensitive medical topics can be particularly advantageous.

References

1. J. Weizenbaum, ELIZA—A computer program for the study of natural language communication between man and machine. Commun. ACM **9**(1), 36–45 (1966)
2. S. Weisband, S. Kiesler, Self disclosure on computer forms: Meta-analysis and implications, in *Proceedings of the SIGCHI Conference on Human Factors in Computing Systems*, (1996), pp. 3–10
3. B. Liu, S. Sundar, Should machines express sympathy and empathy? Experiments with a health advice chatbot. Cyberpsychol. Behav. Soc. Netw. **21**(10), 625–636 (2018)
4. M. Valizadeh, P. Ranjbar-Noiey, C. Caragea, N. Parde, Identifying medical self-disclosure in online communities, in *Proceedings of the 2021 Conference of the North American Chapter of the Association for Computational Linguistics: Human Language Technologies*, (2021), pp. 4398–4408
5. J. Terdiman, Embarrassment is a major barrier to colon cancer prevention, especially among women: A call to action. Gastroenterology **130**(4), 1364–1365 (2006)
6. A. Vaswani, N. Shazeer, N. Parmar, J. Uszkoreit, L. Jones, A.N. Gomez, et al., I.: Attention is all you need. Adv. Neural Inf. Proces. Syst. **30** (2017)
7. J. Devlin, M.W. Chang, K. Lee, K. Toutanova, Bert: Pre-training of deep bidirectional transformers for language understanding. arXiv:1810.04805 (2018)
8. W.H.S. Tsai, D. Lun, N. Carcioppolo, C.H. Chuan, Human versus chatbot: Understanding the role of emotion in health marketing communication for vaccines. Psychol. Mark. **38**(12), 2377–2392 (2021)

Chapter 3
An Enhanced White Shark Optimization Algorithm for Unmanned Aerial Vehicles Placement

Amylia Ait Saadi, Assia Soukane, Yassine Meraihi, Asma Benmessaoud Gabis, Amar Ramdane-Cherif, and Selma Yahia

3.1 Introduction

Over the past decade, the area of Unmanned Aerial Vehicles (UAVs) has experienced significant growth in the commercial, civilian, and military markets [1, 2]. This is primarily due to the tremendous mobility, autonomy, communication, and relatively low cost of UAVs. Therefore, manufacturers today work on fitting and embedding technologies to make UAVs more valuable and suitable for various missions. One of the most promising applications is the application of UAVs to offer various services to connected users in wireless networks. The main purpose of this application's challenging issue is to find the optimal position of UAVs that cover the maximum number of users while ensuring access to the network by connecting the maximum number of drones [3]. The problem of UAVs placement belongs to the group of NP-hard problems successfully solved and optimized by meta-heuristic

A. A. Saadi (✉)
LIST Laboratory, University of M'Hamed Bougara Boumerdes, Boumerdes, Algeria

LISV Laboratory, University of Paris-Saclay, Velizy, France
e-mail: a.aitsaadi@univ-boumerdes.dz

A. Soukane
ECE Paris School of Engineering, Paris, France

Y. Meraihi · S. Yahia
LIST Laboratory, University of M'Hamed Bougara Boumerdes, Boumerdes, Algeria

A. B. Gabis
Ecole nationale Supérieure d'Informatique, Laboratoire des Méthodes de Conception des Systèmes, Alger, Algeria

A. Ramdane-Cherif
LISV Laboratory, University of Paris-Saclay, Velizy, France

M. D. Hina et al. (eds.), *Future Research Directions in Computational Intelligence*, EAI/Springer Innovations in Communication and Computing,
https://doi.org/10.1007/978-3-031-34459-6_3

algorithms. Therefore, in this context, various research based on meta-heuristics were conducted. Authors in [4] applied Elephant Herding Optimization (EHO) algorithm for maximum coverage under different numbers of drones. According to simulation results, EHO performs well in offering maximum coverage for users using a less number of drones. In [5], Ozdag et al. proposed four approaches (OFSAC-PSO, OFSAC-EML, OFSAD-PSO, and OFSAD-EML) based on Particle Swarm Optimization (PSO) and Electromagnetism-Like (EML) algorithms for improving the UAV placement. The performance of the proposed approaches was assessed under different distributions of users and evaluated based on different metrics such as fitness function values, coverage rates, drones altitudes, and 3D drones' locations. Results showed that the OFSAC-PSO algorithm outperforms other optimization methods. In the work of Chaalal et al. [6], Social Spider Optimization (SSO) Algorithm was applied to solve the UAV deployment problem. The effectiveness of SSO algorithm was assessed in three different areas serving different numbers of users and compared to random search (RS) method and uniform distribution application. Simulation results proved that the SSO algorithm outperforms other meta-heuristics in terms of fitness value, execution time, and covered users. Reina et al. [7] proposed a multi-layout multi-sub-population genetic algorithm (MLMPGA) to enhance the UAVs placement for maximum coverage and connectivity. The MLMPGA algorithm was evaluated in various scenarios with different numbers of drones and users and compared to Genetic Algorithm (GA), PSO, and Hill Climbing algorithm (HCA). Test results showed that the MLMPGA algorithm gives competitive results compared to state-of-the-art meta-heuristics regarding the fitness value, coverage, connectivity, and redundancy.

This chapter proposes an ameliorated version of WSO algorithm, called EWSO, based on the incorporation of Elite opposition-based scheme for solving the UAVs deployment problem. The proposed EWSO is tested using 23 cases with various numbers of UAVs and users in comparison with the classical WSO, GWO, and BA algorithms.

The remaining of this chapter is organized as follows. Section 3.2 gives the formulation of the UAVs deployment problem. Section 3.3 gives the description of the WSO algorithm and EOBL strategy. Section 3.4 explains the structure of the proposed EWSO algorithm for solving the UAVs placement issue. Section 3.5 discusses the simulation findings. Finally, Sect. 3.5 shows the concluding remarks and future works.

3.2 UAV Placement Network Model and Problem Formulation

Consider a network system consisting of G users, $G = \{g_1, g_2, \ldots, g_m\}$. These users is served by a set of UAVs $U = \{u_1, u_2, \ldots, u_n\}$ for various applications. To ensure communication and reliability, each UAV is equipped with a radio interface with a maximum transmission range of R_{max} to communicate with ground users and other UAVs. UAVs can take any position defined as (x_j, y_j, h_j), $j \in \{1, 2, \ldots, n\}$ in $3D$ area of dimension $W \times L \times H$. The UAV height is limited by lower and upper bounds h_{min} and h_{max}, respectively. h_{min} is fixed by the user according to the application to protect UAVs from ground threats. h_{max} is related to coverage radius and visibility angle. Let us assume that the users are randomly located at a fixed position (x_i, y_i), $i \in \{1, 2, \ldots, m\}$, where $(x_i, y_i) \in W \times L$. Each user is equipped with a radio interface with a maximum transmission range of R_{max} to communicate with UAV.

The main objective is to find the best UAVs location for maximum user coverage and connectivity, which can be formulated mathematically by the following equation:

$$f(p_i) = \omega_1.\frac{Cv}{m} + \omega_2.\frac{Cn}{n}, \tag{3.1}$$

where Cv denotes the user coverage cost. Cn expresses the UAV connectivity cost. ω is the linear weight coefficient in the range [0, 1], so that $\sum_{i=1}^{2} w_i = 1$.

User Coverage Cost (Cv)

In this chapter, we consider that the users are static in the study area. Each UAV covers a number of users with cover radius defined in Eq. (3.2). We say that the user g_i is covered by the UAV u_j only if the distance between them $d(g_i, u_j)$, expressed in Eq. (3.4), is less than the coverage radius r_j. It is mathematically formulated in Eq. (3.3).

$$r_j = h_j.\tan\left(\frac{\theta}{2}\right), \tag{3.2}$$

where h_j is the UAV height. θ represents the visibility angle.

$$d(g_i, u_j) \leq r_j, \tag{3.3}$$

where $d(g_i, u_j)$ stands for the Euclidean distance between the UAV u_j and the user g_i. It is expressed in the equation below:

$$d(g_i, u_j) = \sqrt{(x_i - x_j)^2 + (y_i - y_j)^2}. \tag{3.4}$$

The total coverage cost by all UAV Cv is represented in the following equation:

$$Cv = \sum_{j=1}^{n} C_{g,u}, \tag{3.5}$$

where $c_{g,u}$ refers to the coverage cost of users by the UAV u_j. To make sure that one user is covered by exactly one UAV, the coverage cost is formulated as follows:

$$C_{g,u} = \begin{cases} 1, & if \min \{d(g, u)\}, \forall u \in U \\ 0 & \text{otherwise}. \end{cases} \tag{3.6}$$

UAV Connectivity Cost (*Cn*)
The main goal of UAVs wireless network is to provide access to different available services. The network is formed by connecting the maximum number of UAVs in a mesh topology to ensure redundancy and availability. We say that the UAV u_j is connected to the UAV u_j only if the distance between them $d(u_j, u_k)$ does not exceed twice the maximum transmission range R_{max}. The connectivity is mathematically represented in Eq. (3.7).

$$Cn = \sum_{j=1}^{n} N_{u_j}, \tag{3.7}$$

where N_{u_j} represents the number of connected UAVs with the UAV u_j in a single hop that is calculated as follows:

$$N_{u_j} = |u_k | d(u_j, u_k) < 2.R_{max}|, \tag{3.8}$$

where $d(u_j, u_k)$ represents the distance between UAVs u_j and u_k, which is expressed in Eq. (3.9).

$$d(u_j, u_k) = \sqrt{(x_j - x_k)^2 + (y_j - y_k)^2 + (h_j - h_k)^2}. \tag{3.9}$$

3.3 Preliminaries

This section describes the definition and concept of WSO algorithm and EOBL strategy.

3.3.1 White Shark Optimizer Algorithm

White Shark Optimizer (WSO) Algorithm is a newly meta-heuristic proposed by Braik et al. [8] in May 2022 for solving optimization problems. The WSO algorithm is a swarm intelligence meta-heuristic that mimics the behavior of the White Shark in hunting preys [8, 9] that can be summarized in three different actions as follows: (I) Movement toward the prey; (II) Random search for the prey; (III) Nearby location of the prey:

(I) Movement toward the prey:

In this behavior, the white shark tracks and locates the prey based on their senses. As a prey moves, a white shark hears wave hesitations that pinpoint the location of its prey and moves directly toward it. This behavior is represented mathematically in Eq. (3.10).

$$v_{t+1}^{i} = \mu \left\{ v_t^i + w_1 \times c_1 \times \left(p_{gbest_t} - p_t^i \right) + w_2 \times c_2 \times \left(p_{best}^{v_t^i} - p_t^i \right) \right\}, \tag{3.10}$$

where v_{t+1}^{i} represents the i-th shark's velocity at $(t + 1)$ iteration $t + 1$. v_t^i is the current velocity of the i-th shark. p_{gbest_t} and p_t^i are the shark's best position and current i-th shark position at iteration t, respectively. $p_{best}^{v_t^i}$ stands for the best known position so far. v_t^i is the current i-th index vector of the white sharks reaching the best position, which is defined in Eq. (3.11). w_1 and w_1 are control parameters represented in Eqs. (3.12) and (3.13). c_1 and c_1 are two random variables. μ represents the constriction factor that is formulated in (3.14).

$$v = [n \times rand(1, n)] + 1, \tag{3.11}$$

where n donates the population size.

$$w_1 = w_{max} + (w_{max} - w_{min}) \times e^{-\left(\frac{4t}{T}\right)^2} \tag{3.12}$$

$$w_2 = w_{min} + (w_{max} - w_{min}) \times e^{-\left(\frac{4t}{T}\right)^2}, \tag{3.13}$$

where t and T are the current and the maximum number of iterations, respectively. w_{min} and w_{max} stand for the initial and subordinate velocities, respectively.

$$\mu = \frac{2}{|2 - \alpha - \sqrt{\alpha^2 - 4\alpha}|}, \tag{3.14}$$

where α is a fixed value at 4.125 that represents the acceleration coefficient.

(II) Random search for the prey:

In this case, the white shark follows the prey tracks based on the smelling and hearing in random positions. This movement is mathematically formulated by the following equation:

$$p_{t+1}^{i} = \begin{cases} p_t^i . \neg \oplus p_0 + ub.a + lb.b, & rand < ws \\ p_t^i + \frac{v_t^i}{f}, & rand \geq ws \end{cases}, \tag{3.15}$$

where p_{t+1}^{i} denotes the new position of the i-th white shark. $\neg$ is a negation operation. p_0 refers to a logical position vector defined in Eq. (3.18). ub and lb represent the upper and lower search space boundaries, respectively. C is a control parameter that balances the exploration and exploitation, which is expressed in Eq. (3.20). f stands for the frequency of the white shark's wavy motion that can be calculated as shown in Eq. (3.19).

$$a = sgn\left(p_t^i - ub\right) > 0 \tag{3.16}$$

$$b = sgn\left(p_t^i - lb\right) < 0 \tag{3.17}$$

$$p_0 = \oplus(a, b), \tag{3.18}$$

where $\oplus$ represents the bit-wise exclusive-or (XOR) operator.

$$f = f_{min} + \frac{f_{max} - f_{min}}{f_{max} + f_{min}}, \tag{3.19}$$

where f_{min} and f_{max} refer to the minimum and maximum frequencies of the white shark's wavy motion, respectively.

$$C = \frac{1}{\alpha_0 + e^{\left(\frac{T}{2} - t\right)\alpha_1}}, \tag{3.20}$$

where α_0 and α_1 are represented as two positive values to manage both exploration and exploitation behavior.

(III) Nearby location of the prey: In this method, the white shark uses a fish school technique and moves toward the shark that is closer to the prey. This movement is formulated by the following equation:

$$p_{t+1}^{\prime i} = p_{gbest_t} + r_1.\overrightarrow{D_p}.sgn(r_2 - 0.5), \quad r_3 < s, \tag{3.21}$$

where $p_{t+1}^{\prime i}$ is the updated i-th white shark's position. r_1, r_2, and r_3 are random variables in the range of [0, 1]. $\overrightarrow{D_p}$ represents the distance between the prey and the white shark. $sgn(r_2 - 0.5)$ is a parameter used to change the search direction. s expresses white shark's senses that are presented in Eq. (3.23).

$$\overrightarrow{D_p} = |rand.\left(p_{gbest_t} - p_t^i\right)|, \tag{3.22}$$

where $rand$ is a random variable within the range of [0, 1]. p_t^i stands for the current i-th white shark's position.

$$s = |1 - e^{\left(\frac{-\alpha_2 . t}{T}\right)}|, \tag{3.23}$$

where α_2 represents the control behavior parameter.
White shark's position update according to fish school behavior is given in Eq. (3.24).

$$p_{t+1}^i = \frac{p_t^i - p_{t+1}^{'i}}{2.rand}. \tag{3.24}$$

3.3.2 *Elite Opposition-Based Learning*

Opposition-based learning strategy (OBL) proposed by Tizhoosh [10] is a well-regarded intelligent strategy that aims to enhance the chance of finding more effective solution by checking simultaneously the initial solution and its corresponding opposite solution. Let us consider a given candidate solution p in one dimension search space delimited by $[Lb, Ub]$. Then, the opposite solution $\overline{p}$ is defined as follows:

$$\overline{p} = Ub + Lb - p. \tag{3.25}$$

Elite opposition-based learning (EOBL) is an improved version of OBL widely combined with several meta-heuristics. The basic concept of EOBL is to employ first an elite solution that has expectantly more information than other individuals and then generate the opposite of the current solution in the search area. The elite individual leads the population toward the promising area where the global solution can be found. The elite opposite solution can be formulated by the following expression:

$$\overline{p_i} = r.(Du_j + Dl_j) - p_j, \quad j = 1, \ldots, Dim, \tag{3.26}$$

where r is a random number in the range $[0 - 1]$. Du_j and Dl_j are dynamic boundaries that can be presented as follows:

$$Du_j = max(p_j), \quad Dl_j = min(p_j). \tag{3.27}$$

However, the elite solution may jump out of the search space boundaries $[Lb, Ub]$. Consequently, EOBL will fail to consider a valid solution. To overcome this issue, we address a random value for this kind of solution as follows:

$$\overline{p_i} = rand(Lb_j, Ub_j) \quad if \quad \overline{p_j} < Lb_j \parallel \overline{p_j} > Ub_j. \tag{3.28}$$

3.4 Elite Opposition-Based White Shark Optimization Algorithm for UAVs Placement

This section describes the implementation steps of our proposed EWSO algorithm for solving the UAVs placement problem. In this sense, the EOBL strategy was incorporated into WSO to enhance its optimization performance. The proposed EWSO involves mainly four steps including initialization, evaluation, update, and finally termination that displays the best UAVs positions found.

3.4.1 *Initialization*

The first step of implementing the proposed EWSO algorithm consists of initializing white sharks positions randomly in the search area that is bounded by the size of the deployment area for UAVs. The initial position of White sharks is represented in a $N \times D$ matrix as shown in Eq. (3.29).

$$\text{Positions} = \begin{bmatrix} Pos_{1,1} & Pos_{1,2} & \dots & Pos_{1,D} \\ Pos_{2,1} & Pos_{2,2} & \dots & Pos_{2,D} \\ \dots & \dots & \dots & \dots \\ Pos_{N,1} & Pos_{N,2} & \dots & Pos_{N,D} \end{bmatrix}, \tag{3.29}$$

where N and D stand for the population size and the problem dimension, respectively. Pos represents the i-th shark position that expresses the UAVs positions. It is formulated in the following equation:

$$Pos_i = (\{x_{i,1}, x_{i,2}, \dots, x_{i,n}\}; \{y_{i,1}, y_{i,2}, \dots, y_{i,n}\}; \{z_{i,1}, z_{i,2}, \dots, z_{i,n}\}). \tag{3.30}$$

3.4.2 *Evaluation*

According to this population, the EWSO algorithm applies an evaluation by calculating the fitness value of the current population. Based on this evaluation, EWSO finds the initial best position that corresponds to the maximum fitness value as formulated in Eq. (3.31), considering the UAVs placement problem is a maximization problem.

$$Pos_{best} = \arg\max f(Pos). \tag{3.31}$$

After evaluation, the EWSO initializes the dynamic bounds $[Dl, Du]$ using Eq. (3.27) to process the next step.

Algorithm 1 Elite opposition-based white shark optimization algorithm for UAVs placement

Initialize EWSO parameters: Maximum number of iterations T, Population' size N, Dimension Dim, μ, v, f, $etc.$
Initialize the population of EWSO: $Pos_i (i = 1, 2, ..., N)$
Calculate the fitness value $f(Pos_i)$
Determine the best position Pos_{best}
while $(t < T)$ **do**
  for $i = 1, 2, \ldots, N$ **do**
    Find the N opposite positions based on EOBL using Eqs. (3.27) and (3.26), and select the N fittest positions using Eq. (3.32)
    $v_{t+1}^i = \mu\{v_t^i + p_1 \times c_1 \times (p_{gbest_t} - p_t^i) + p_2 \times c_2 \times (p_{best}^{v_t^i} - p_t^i)\}$
    Update the position using Eq. (3.15)
    if $rand < s$ **then**
      Update the position using Eq. (3.21)
      Update the final position using Eq. (3.24)
    end if
    Update the best position Pos_{best}
  end for
  $t = t + 1$
end while
return The best position Pos_{best}

3.4.3 Update

In this step, the EWSO algorithm searches for the elite opposite of the current population $\overline{Pos_i}$ using Eq. (3.26). Then, EWSO evaluates both Pos_i and $\overline{Pos_i}$ and selects N best candidates for updating according to the best fitness value as represented in Eq. (3.32).

$$Pos_i = \arg\max f(Pos_i \cup \overline{Pos_i}). \tag{3.32}$$

After positions' selection, the EWSO processes as original WSO for updating position by using Eqs. (3.10), (3.15), (3.21), and (3.24). The new positions are evaluated, and the dynamic bounds are updated according to the new positions. The EWSO algorithm repeats this step until the maximum number of iterations is reached.

3.4.4 Termination

This step represents the end of the process. The EWSO displays the best UAVs positions found represented in an array as follows:

$$Pos_{best} = (\{x_1, x_2, \ldots, x_n\}; \{y_1, y_2, \ldots, y_n\}; \{z_1, z_2, \ldots, z_n\}). \tag{3.33}$$

The pseudocode of our proposed EWSO algorithm for UAVs placement can be summarized in Algorithm 1.

3.5 Numerical Results

In this section, the evaluation of the proposed EWSO algorithm for solving the UAV placement is presented. The evaluation is done on several experiments and configurations as shown in Table 3.1 and compared to BA [11], GWO [12], and the original WSO algorithm. All simulations are running using MATLAB 2021B Software installed on Core i7 2.90 GHz, RAM 32 GB machine.

The effectiveness of the EWSO algorithm was evaluated by considering the fitness value, user coverage, and UAVs connectivity. The reported results represent the average of 50 runs for each metric found by each algorithm.

3.5.1 Impact of Varying the Number of UAVs

In the first scenario, the number of UAVs is varied from 4 to 24 with a step of 2, and the number of users is fixed at 200. The obtained results, in this case, are reported in Figs. 3.1, 3.2, and Table 3.2. We can clearly notice that the number of UAVs is proportional to the quality provided. Increasing the number of drones increases the fitness value. The EWSO algorithm reaches the highest fitness value of 0.9715 when the number of UAVs is fixed at $U = 20$. With 20 UAVs, EWSO covers more than 90% of users with 99.9% of connectivity, while BA, GWO, and WSO obtained the maximum fitness value at $U = 22$ and $U = 24$, respectively. Significantly, EWSO requires fewer UAVs than BA, GWO, and WSO to achieve the highest quality, positively impacting cost and energy.

Table 3.1 Description of experiences

Scenario parameters	Value
Population size	50
Maximum number of iterations	200
Weight coefficient w	$[\{0.5, 0.5\}]$
Number UAVs	[4–24]
Number of users	[50–300]
Maximum transmission range	100 m
Visibility angle	120
Area dimension	1000 m × 1000 m

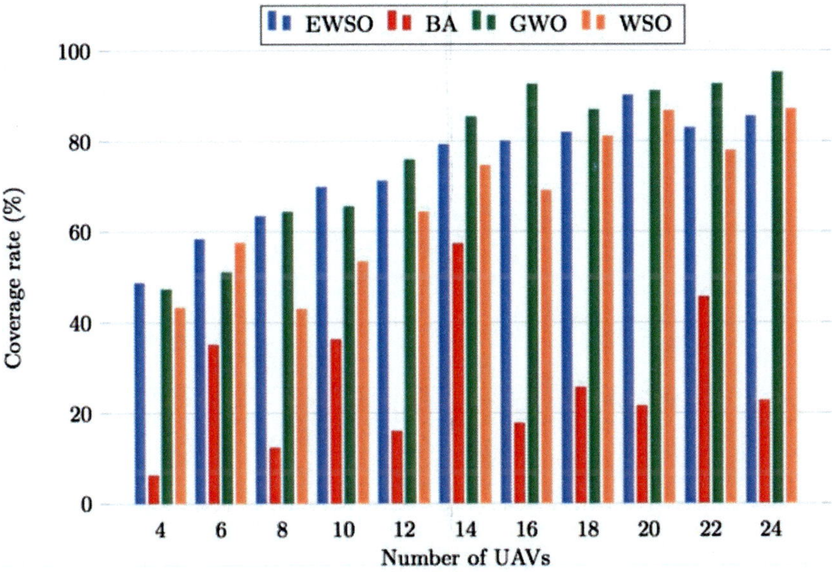

Fig. 3.1 The coverage rate using a different number of drones

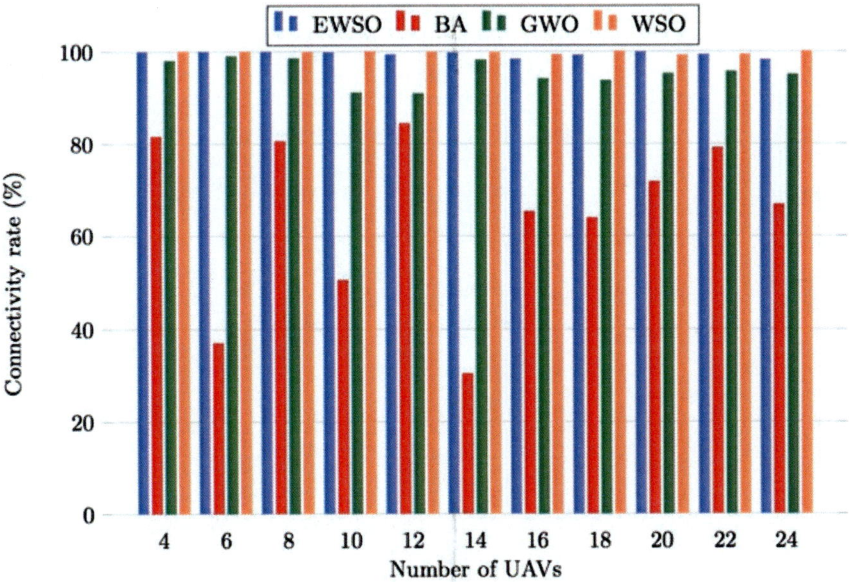

Fig. 3.2 The connectivity rate using a different number of drones

Table 3.2 Results obtained from different algorithms in the first case

Results	EWSO	BA	GWO	WSO
	$U = 4$			
Fitness	0.7617	0.4386	0.7264	0.7223
Coverage (%)	48.71	6.23	47.28	43.19
Connectivity (%)	100	81.5	98	100
	$U = 6$			
Fitness	0.8107	0.3602	0.7502	0.8024
Coverage (%)	58.31	35.04	51.03	57.47
Connectivity (%)	100	37	99	100
	$U = 8$			
Fitness	0.8336	0.4639	0.8172	0.7294
Coverage (%)	63.38	12.27	64.93	42.87
Connectivity (%)	100	80.5	98.5	99.75
	$U = 10$			
Fitness	0.8651	0.4327	0.7828	0.786
Coverage (%)	69.84	36.18	65.57	53.36
Connectivity (%)	99.8	50.4	91	100
	$U = 12$			
Fitness	0.8723	0.5013	0.8335	0.8347
Coverage	71.22	15.92	75.86	64.36
Connectivity	99.33	84.33	90.83	99.83
	$U = 14$			
Fitness	0.9173	0.4373	0.917	0.8904
Coverage (%)	79.21	57.2857	85.26	74.58
Connectivity (%)	99.71	30.29	98.14	99.71
	$U = 16$			
Fitness	0.9224	0.4145	0.9328	0.8643
Coverage (%)	79.98	17.65	92.44	69.06
Connectivity (%)	98.37	65.25	94.12	99.25
	$U = 18$			
Fitness	0.9316	0.4422	0.9025	0.922
Coverage (%)	81.87	24.55	86.84	81.01
Connectivity (%)	99.22	63.89	93.67	99.98
	$U = 20$			
Fitness	0.9715	0.4657	0.9306	0.9459
Coverage (%)	90.06	21.44	90.99	86.62
Connectivity (%)	99.9	71.7	95.2	99.1
	$U = 22$			
Fitness	0.9371	0.6225	0.9407	0.9106
Coverage (%)	82.36	45.5	92.5	77.8
Connectivity (%)	99.27	79	95.63	99.27
	$U = 24$			
Fitness	0.9512	0.4716	0.9508	0.9491
Coverage (%)	85.41	22.57	95.15	86.95
Connectivity (%)	98.17	66.75	95	99.91

3.5.2 *Impact of Varying the Number of Users*

In this scenario, the number of UAVs is fixed at 20, and the number of users varies from 50 to 300 with a step of 25. Figures 3.3, 3.4, and Table 3.3 describe the results

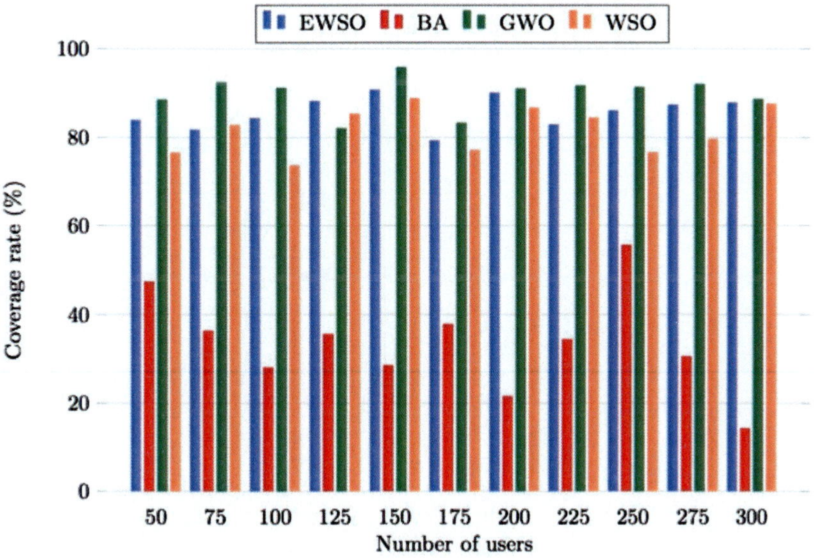

Fig. 3.3 The coverage rate using a different number of users

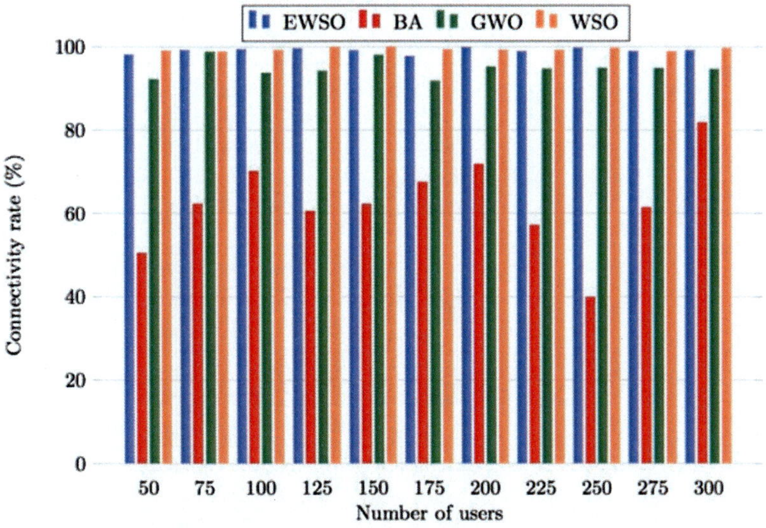

Fig. 3.4 The connectivity rate using a different number of drones

Table 3.3 Results obtained from different algorithms in the second test

Results	EWSO	BA	GWO	WSO
	$G = 50$			
Fitness	0.9476	0.4884	0.9029	0.9196
Coverage	83.88	47.28	88.48	76.44
Connectivity	98.1	50.4	92.1	99
	$G = 75$			
Fitness	0.9345	0.4928	0.9261	0.9423
Coverage	81.73	36.26	92.42	82.72
Connectivity	99.2	62.3	98.8	98.8
	$G = 100$			
Fitness	0.9435	0.4909	0.9242	0.8912
Coverage	84.34	27.98	91.14	73.48
Connectivity	99.4	70.1	93.7	99.1
	$G = 125$			
Fitness	0.9667	0.4804	0.8812	0.9476
Coverage	88.21	35.48	82.05	85.28
Connectivity	99.6	60.5	94.2	99.9
	$G = 150$			
Fitness	0.9767	0.4533	0.969	0.9602
Coverage	90.72	28.45	95.8	88.72
Connectivity	99.1	62.2	98	99.9
	$G = 175$			
Fitness	0.923	0.5255	0.8784	0.8983
Coverage	79.29	37.6914	83.24	76.97
Connectivity	97.7	67.4	91.7	99.2
	$G = 200$			
Fitness	0.9715	0.4657	0.9306	0.9459
Coverage (%)	90.06	21.44	90.99	86.62
Connectivity (%)	99.9	71.7	95.2	99.1
	$G = 225$			
Fitness	0.9343	0.4571	0.9321	0.9366
Coverage	82.88	34.3228	91.72	84.37
Connectivity	98.9	57.1	94.7	99.1
	$G = 250$			
Fitness	0.9549	0.4767	0.9312	0.896
Coverage	86.04	55.5	91.33	76.48
Connectivity	99.7	39.85	94.9	99.6
	$G = 275$			
Fitness	0.9571	0.4591	0.9341	0.9145
Coverage	87.33	30.4291	92.02	79.59
Connectivity	98.9	61.4	94.8	98.8
	$G = 300$			
Fitness	0.9605	0.4794	0.9162	0.9498
Coverage	87.86	14.1733	88.64	87.55
Connectivity	99.1	81.7	94.6	99.6

obtained for different numbers of users covered by 20 UAVs. As the number of users increased, fitness value, coverage, and connectivity metrics increased until $G = 150$, where all algorithms except BA reached maximum performance. Therefore, the optimal number of UAVs to cover 150 users using this configuration is 20. For other user cases, solutions provided by EWSO, GWO, and WSO algorithms are good and acceptable.

In most cases, the proposed EWSO algorithm outperforms the others by giving the highest fitness results that reflect the best balance between the coverage and the connectivity objectives.

3.6 Conclusion

In this chapter, we have proposed an ameliorated version of WSO, named EWSO, for tackling the problem of UAVs placement in 5G networks. The Elite opposition-based learning strategy is incorporated into the original WSO to enhance its efficiency. The proposed EWSO was tested using 23 scenarios with 24 UAVs and 300 users compared to WSO, GWO, and BA algorithms. The simulation results proved the superiority and efficiency of EWSO considering fitness values, coverage, and connectivity parameters. For future work, several directions in which EWSO can be extended. Initially, the energy consumption problem for UAVs can be addressed in this context. Furthermore, the EWSO algorithm can be enhanced by combining other meta-heuristics for better quality.

References

1. M. Radmanesh, M. Kumar, P.H. Guentert, M. Sarim, Overview of path-planning and obstacle avoidance algorithms for UAVs: a comparative study. Unmanned Syst. **6**(02), 95–118 (2018)
2. S. Aggarwal, N. Kumar, Path planning techniques for unmanned aerial vehicles: a review, solutions, and challenges. Comput. Commun. **149**, 270–299 (2020)
3. I.A. Elnabty, Y. Fahmy, M. Kafafy, A survey on UAV placement optimization for UAV-assisted communication in 5G and beyond networks. Phys. Commun. **51**, 101564 (2022)
4. I. Strumberger, N. Bacanin, S. Tomic, M. Beko, M. Tuba, Static drone placement by elephant herding optimization algorithm, in *2017 25th Telecommunication Forum (TELFOR)* (IEEE, Piscataway, 2017), pp. 1–4
5. R. Ozdag, Multi-metric optimization with a new metaheuristic approach developed for 3D deployment of multiple drone-BSs. Peer-to-Peer Networking Appl. **15**(3), 1535–1561 (2022)
6. E. Chaalal, L. Reynaud, S.M. Senouci, A social spider optimisation algorithm for 3D unmanned aerial base stations placement, in *2020 IFIP Networking Conference (Networking)* (IEEE, Piscataway, 2020), pp. 544–548
7. D.G. Reina, H. Tawfik, S.L. Toral, Multi-subpopulation evolutionary algorithms for coverage deployment of UAV-networks. Ad Hoc Networks **68**, 16–32 (2018)
8. M. Braik, A. Hammouri, J. Atwan, M.A. Al-Betar, M.A. Awadallah, White shark optimizer: a novel bio-inspired meta-heuristic algorithm for global optimization problems. Knowledge-Based Syst. **243**, 108457 (2022)

9. M.A. Ali, S. Kamel, M.H. Hassan, E.M. Ahmed, M. Alanazi, Optimal power flow solution of power systems with renewable energy sources using white sharks algorithm. Sustainability **14**(10), 6049 (2022)
10. H.R. Tizhoosh, Opposition-based learning: a new scheme for machine intelligence, in *International Conference on Computational Intelligence for Modelling, Control and Automation and International Conference on Intelligent Agents, Web Technologies and Internet Commerce (CIMCA-IAWTIC'06)*, vol. 1 (IEEE, Piscataway, 2005), pp. 695–701
11. X.-S. Yang, A. Hossein Gandomi, Bat algorithm: a novel approach for global engineering optimization. Eng. Comput. **29**(5), 464–483 (2012)
12. S. Mirjalili, S.M. Mirjalili, A. Lewis, Grey wolf optimizer. Adv. Eng. Software **69**, 46–61 (2014)

Chapter 4
Computer-Aided Diagnosis Based on DenseNet201 Architecture for Psoriasis Classification

Abdelhak Mehadjbia, Khadidja Belattar, and Fouad Slaoui Hasnaoui

4.1 Introduction

The automatic skin disease diagnosis is considered of great importance in the medical field. Psoriasis is a systemic, chronic, genetically determined, progressive and inflammatory disorder of the skin, joints, and nails. It is characterized by the excessive proliferation of epidermal cells and immune inflammation, resulting in the formation of sharply demarcated, scaly, and erythematous plaques [18]. Such symptoms negatively impact the life quality of the patient and cause significant physical and psychosocial burden [15]. When diagnosing the disease, the clinical manifestations and the histological clues should be considered. The morphology, the distribution, the severity, and the course of a suspected psoriasis serve as the basis of the clinical diagnosis of the skin condition. The histological findings consist of hyperkeratosis, parakeratosis, acanthosis of the epidermis with dilated blood vessels and a lymphocytic infiltrate [13].

Several kinds of psoriasis are recognized. We can distinguish five major variants, namely: plaque, guttate, pustular, inverse, and erythrodermic psoriasis [14]. Also, it is possible to have a mild, moderate, or severe form of the condition depending on several factors, importantly: the patient quality of life, the coverage, and the location and appearance of the psoriasis.

Whether the psoriasis is mild, moderate, or severe, it could develop further related comorbidities and complications. These include: psoriatic arthritis, anxiety,

---

A. Mehadjbia (✉) · K. Belattar
Computer Science Department, University of Algiers, Alger Ctre, Algeria
e-mail: k.belattar@univ-alger.dz

F. S. Hasnaoui
Université du Québec en Abitibi-Témiscamingue, Rouyn-Noranda, QC, Canada
e-mail: fouad.slaoui-hasnaoui@uqat.ca

M. D. Hina et al. (eds.), *Future Research Directions in Computational Intelligence*, EAI/Springer Innovations in Communication and Computing,
https://doi.org/10.1007/978-3-031-34459-6_4

depression, obesity, hypertension, diabetes mellitus, hyperlipidemia, metabolic syndrome, smoking, cardiovascular disease, alcoholism, Crohn's disease, lymphoma, and multiple sclerosis [12]. Therefore, an early diagnosis and consistent monitoring of the psoriasis would help manage the patient's overall health.

Generally speaking, dermatologists primarily use the clinical examination (inspection, palpation, percussion, and auscultation) to diagnose the suspected psoriasis lesion.

However, the current diagnostic method is laborious, confused, not reliable enough and leads to the subjectivity in decision-making. This is mainly owing to: the lack of well-trained dermatologists, the artifact occlusion (such as hair strands), and the high inter-class clinical similarity across papulosquamous skin diseases, namely: nummular eczema, mycosis fungoides, pityriasis rubra pilaris, Duhring's disease, and Bowen's disease [27].

Timely and accurate diagnosis of the psoriasis may raise an important health concern [4]. So, the automatic process of the diagnosis basically involves the image classification, and convolutional neural network (CNN) is one of the methods to deal with such medical image problems [5]. The most widely used CNN architectures are: VGG-16 [23], ResNet [7], InceptionNet [24], Xception [3], DenseNet [11], MobileNet [8], RestNext [29], and SeResNeXt [9].

Our main focus of this chapter is to facilitate and improve the psoriasis differential diagnosis performance by exploiting the state-of-the-art CNN architectures in particular DenseNet201 [10], InceptionResNetV2 [25], and NasNetMobile [32], which they have not been addressed to the best of our knowledge.

The paper is organized as follows. In Sect. 4.2, we provide some background information about the adapted DenseNet201 model. In Sect. 4.3, we review recent works on the psoriasis classification problem. In Sect. 4.4, we present the computer-aided psoriasis diagnosis system. In Sect. 4.5, we summarize and discuss the experimental results conducted on the public psoriasis image database. In Sect. 4.6, we provide concluding remarks and future research directions.

4.2 The Transfer-Learning-Based DenseNet201 Model

In this chapter, we used DenseNet201 model for the psoriasis image classification. The model is the abbreviation for densely connected convolutional networks. Several versions of DenseNet have been proposed, including: DenseNet121, DenseNet169, DenseNet201, and DenseNet264. Each version consists of four dense blocks with a different number of layers. The overall structure of the DenseNet is illustrated in Fig. 4.1. According to the figure, DenseNet201 is composed of:

- Convolution and max pooling layers.
- Dense blocks: are aimed to reduce the spatial dimensions and the number of the input feature maps. Each dense block has two convolution operations of 1×1

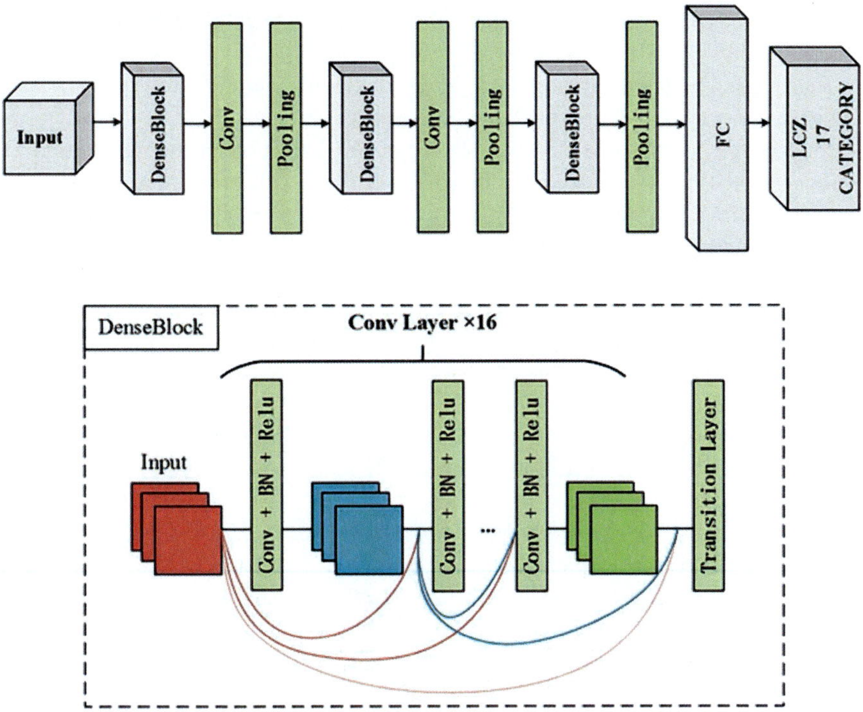

Fig. 4.1 The basic architecture of DenseNet201

and 3 × 3 kernels size. In a single block, each layer is connected to all subsequent layers.
- Transition layers: are used between the dense blocks, consist of a batch normalization layer, 1 × 1 convolution followed by a 2 × 2 average pooling layer.
- Fully connected (FC) block for the prediction.

4.3 Related Work

To summarize the relevant works related to the psoriasis classification using deep learning models, we searched the recently published journal and conference articles from several scientific databases. During the reviewing process, we found several research studies in the literature that addressed the classification of pigmented skin tumors. A comprehensive survey of the skin disease diagnosis with deep learning techniques can be found in [16, 17] and [22].

To the best of our knowledge, only a few related papers were retrieved for the classification and severity assessment of the psoriasis using deep learning models.

This is mainly due to the lack of available large-scale labeled psoriasis image datasets.

Starting with deep learning techniques, previous literature for the psoriasis diagnosis includes the approach proposed by Zhang et al. [31]. The authors introduced the human knowledge into GoogleNet InceptionV3 model to improve the diagnosis of four types of cutaneous diseases (basal cell carcinoma, melanocytic nevus, seborrheic keratosis, and psoriasis). The proposed approach achieved an average accuracy ($\pm$ standard deviation) of 87.25 $\pm$ 2.24% on the test dataset with 1067 images.

Another work in that class of approaches is that of Peng et al. [21] who presented the classification model of the psoriasis based on deep residual network (ResNet-34). The used model was evaluated on 30,000 images of five classes: four types of psoriasis and one class of normal images. The ResNet-34 model outperforms VGG model in terms of the recall rate, F1 score, and area under curve (AUC) score.

Meanwhile, Yiguang et al. [30] adapted EfficientNet-B4 model to diagnose the dermoscopic psoriasis images. To evaluate the model, the authors used a collection of 7033 dermoscopic images of five kinds of papulosquamous skin diseases, namely: psoriasis, eczema, lichen planus, pityriasis rosea, and seborrheic dermatitis. The used EfficientNet-B4 model achieved 0.929 sensitivity and a specificity of 95.2% for diagnosing psoriasis, 77.3% sensitivity, and a specificity of 92.6% for diagnosing eczema, 0.933 sensitivity, and a specificity of 96% for diagnosing lichen planus, and 84% sensitivity and a specificity of 98.5% for diagnosing other classes. As a result, the performance of the EfficientNext-B4 model was comparable to that of dermatologists.

In Aijaz et al. [1], the authors used two deep learning neural networks for the psoriasis classification, namely: CNN and long short-term memory (LSTM). To evaluate the stated models, the authors used a collection of 473 images belonging to five types of psoriasis (plaque, guttate, inverse, pustular, and erythrodermic). The reported accuracies of CNN and LSTM are 84.2% and 72.3%, respectively.

Besides the conventional machine learning approaches, support vector machines (SVMs) model was employed by Umapath et al. [28] to classify the psoriasis and normal hand regions. For this purpose, the gray-level co-occurrence matrix (GLCM) features are extracted from the hand thermal images. The sensitivity, specificity, and accuracy of the classifier were found to be 83.3%, 73.3%, and 78.3%, respectively.

From the previous literature, we could highlight that most of the research studies relies on the CNN models as they are the most suitable for the image classification to learn the most discriminative features. Furthermore, they do not require more computing power in the model training as well as in the classification compared to the recurrent neural networks.

In this chapter, we propose to use the DenseNet201 architecture since it has proven to be superior than many state-of-the-art classification algorithms when adopted to the skin cancer classification context, according to a recent study [2]. In our context, we take advantage of this successful model to perform the present study. Our objective is to facilitate and improve the psoriasis diagnosis.

4.4 The Proposed Computer-Aided Psoriasis Diagnosis System

The computer-aided diagnosis (CAD) system has a significant potential in the field of dermatology. It enables early and an enhanced recognition of the suspected skin lesions, which facilitates the timely interventions and improving the outcomes. The CAD can provide the dermatologists, or even untrained physicians, with a valuable aid for the diagnosis of the skin diseases by classifying the input image according to the specific image features.

The block diagram of the developed computer-aided diagnosis system is shown in Fig. 4.2. It involves two stages: (1) Psoriasis image database building (offline) and (2) Model training and image classification (online) phases.

In the offline phase, the collected psoriasis skin lesion images are prepared for the following stages of the psoriasis image diagnosis. Next, we applied the sampling technique to generate the training, validation, and test image sets.

In the online phase, a test psoriasis query image is classified based on the trained deep neural network model. The following subsections describe the involved stages in the psoriasis diagnosis process.

4.4.1 Psoriasis Image Preparation

To boost the performance of the psoriasis diagnosis system, the images should be well prepared before feeding the used data as input to the deep neural network model. For this purpose, we applied image resizing and image normalization tasks. All images were normalized and re-scaled to 224 × 224 pixels to facilitate the use of the neural network model.

Data augmentation [19] is also established on the prepared dataset (composed of about 479 images) to enlarge it with new samples and robustify the used model against to the real-world scenarios (i.e., perspective changes, different camera perspectives, and contrast and brightness changes).

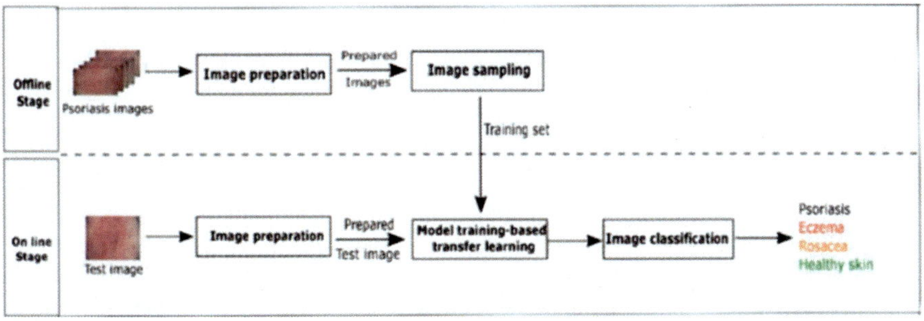

Fig. 4.2 The proposed computer-aided psoriasis diagnosis system

The data augmentation operations include: geometric distortion, brightness adjustment (lightening and darkening), flipping (left, right, top–bottom, and bottom–top), random rotation by 90°,180° or 270°, and shearing. With the data augmentation, we have generated a dataset of 4212 images, containing 1135 psoriasis images, 1068 eczema images, 1002 rosacea images, and 1006 healthy skin images, respectively. After preparing the dataset, the next stage is sampling the dataset.

4.4.2 *Psoriasis Image Sampling*

The basic idea here is to split the dataset into 70% for training, 15% for validation, and 15% withheld for testing such that we have an equal number of samples of each class. Table 4.1 shows the distribution of images among different classes in the training, validation, and test sets for the augmented dataset.

Once the three datasets are generated, the training and classification stages of the adapted CNN architecture are established.

4.4.3 *Deep Neural Network Model Training*

The structure of the adapted DenseNet201 model is exhibited in Fig. 4.3.

To better train the adapted model and make predictions of the psoriasis skin lesion images, transfer learning [20] is involved. During the transfer learning,

Table 4.1 The skin lesion image distribution in the training, validation, and test sets

	Psoriasis	Eczema	Rosacea	Healthy skin	Total
Training set	794	747	701	704	2246
Validation set	170	160	150	150	630
Test set	171	161	151	153	635

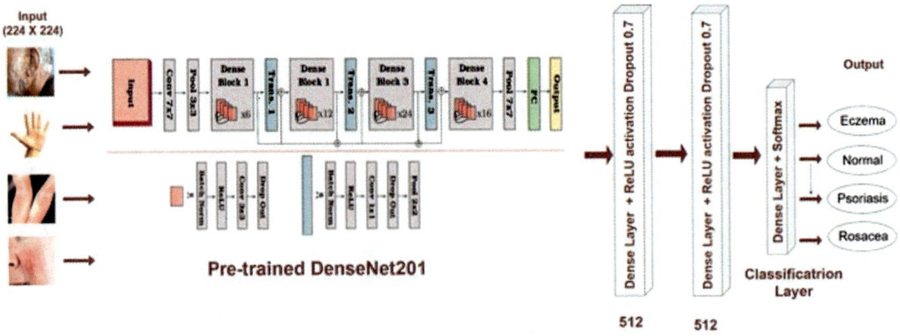

Fig. 4.3 The proposed computer-aided psoriasis diagnosis system

knowledge is leveraged from a source problem in a new related target task. The basic principle on which the transfer learning works is to freeze the early layers of a pre-trained (i.e., DenseNet201) model and fine-tune the rest of its layers on the target (i.e., skin lesion image classification) task.

In the model training process, we used cross-entropy loss function, Adam optimizer, and softmax activation function. We selected these functions according to some conducted tests.

4.4.4 *Psoriasis Image Classification*

After training the adapted DenseNet201 model, we used the generated model to perform the classification of the test psoriasis images. The output class for a given test image could be: psoriasis, eczema, rosacea, or healthy skin classes.

4.5 Experimental Study

The experiments have been carried out on the online Google Colab Pro platform. It provides a free GPU (K80, T4, P4, and P100 of NVidia) with a 25 GB of RAM. The development of the used CNN models has been done using Python programming language.

4.5.1 *Dataset Description*

In this chapter, we collected a dataset from the public DermNet website.

The DermNet skin disease atlas consists of 23 classes of skin diseases, contained over 23,000 labeled images. Since our objective is to establish a diagnosis system of the psoriasis skin lesions, we used a collection of 364 images of three confused kinds of papulosquamous skin diseases, comprising psoriasis, eczema, and rosacea. They show similar clinical symptoms, including rashes, bumps, redness, and itching that may present more differential diagnostic difficulty. Besides, the healthy skin images are included to have a dataset of 479 images (Table 4.2).

4.5.2 *Empirical Parameter Setting*

The objective of the tuning process is to look for the best hyper-parameters values so that the adapted models can have an enhanced discrimination ability. To do so, we performed extensive training of the adapted DenseNet201, InceptionRes-

Table 4.2 The skin lesion image samples

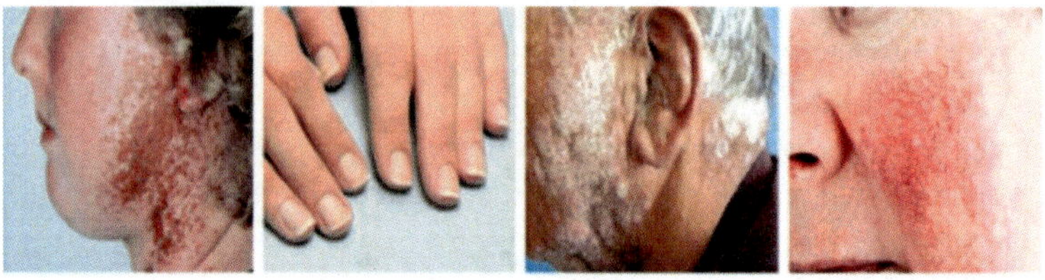

(a) Eczema (b) Healthy skin (c) Psoriasis (d) Rosacea

NetV2, NasNetMobile, ResNet101 [6], EfficientNetB4, and EfficientNetB6 [26] architectures. Hence, we tuned the batch size, the max epoch, and the learning rate hyper-parameters as follows. The max epoch is set to 100, the batch size is chosen to be 16, and the learning rate is 0.0001.

4.5.3 Performance Evaluation

To evaluate the performance of the used CNN models, we computed the classification accuracy, precision, recall, sensitivity, and specificity. They can be defined as follows:

$$\text{Accuracy} = \frac{\text{TP} + \text{TN}}{\text{TP} + \text{TN} + \text{FP} + \text{FN}} \quad (1),$$

$$\text{Precision} = \frac{\text{TP}}{\text{TP} + \text{FP}} \quad (2), \quad \text{Recall} = \frac{\text{TP}}{\text{TP} + \text{FN}} \quad (3),$$

$$\text{Sensitivity} = \frac{\text{TP}}{\text{TP} + \text{FN}} \quad (4), \quad \text{Specificity} = \frac{\text{TN}}{\text{TN} + \text{FP}} \quad (5),$$

where TP, TN, FP, and FN, obtained from the confusion matrix of the corresponding model, denote the number of true positives, the number of true negatives, the number of false positives, and the number of false negatives, respectively.

4.5.4 Results and Discussion

In this section, we present the results of the DenseNet201 model-based psoriasis classification. We also compare the adapted DenseNet201 model against InceptionResNetV2, NasNetMobile, ResNet101, EfficientNetB4, and EfficientNetB6 architectures. Figure 4.4 illustrates the accuracy and loss progress through epochs for the above-mentioned models.

From the DenseNet201 curves, we observe that the initial validation loss is above 1.00, but after ten epochs the loss decreases below 0.5. In the same manner, in the accuracy model, the initial validation accuracy is below 0.6, but after one epoch, the validation accuracy suddenly increases to nearly 0.83. So, there is a positive trend toward improving the classification accuracy and reducing the loss. The model converges in epoch 100 with an accuracy of almost 92% and a loss of 0.25.

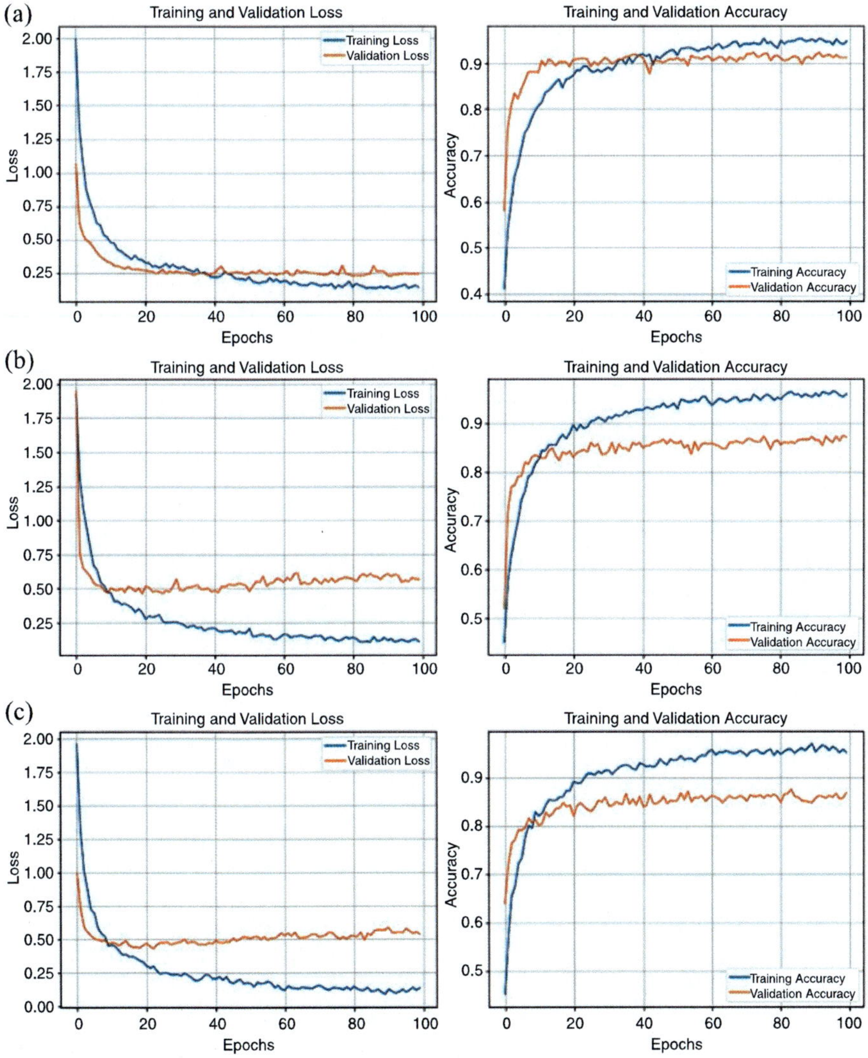

Fig. 4.4 Loss and accuracy curves of the adapted CNN architectures. (**a**) DenseNet201. (**b**) InceptionResNetV2. (**c**) NasNetMobile. (**d**) ResNet101. (**e**) EfficientNetB4. (**f**) EfficientNetB6

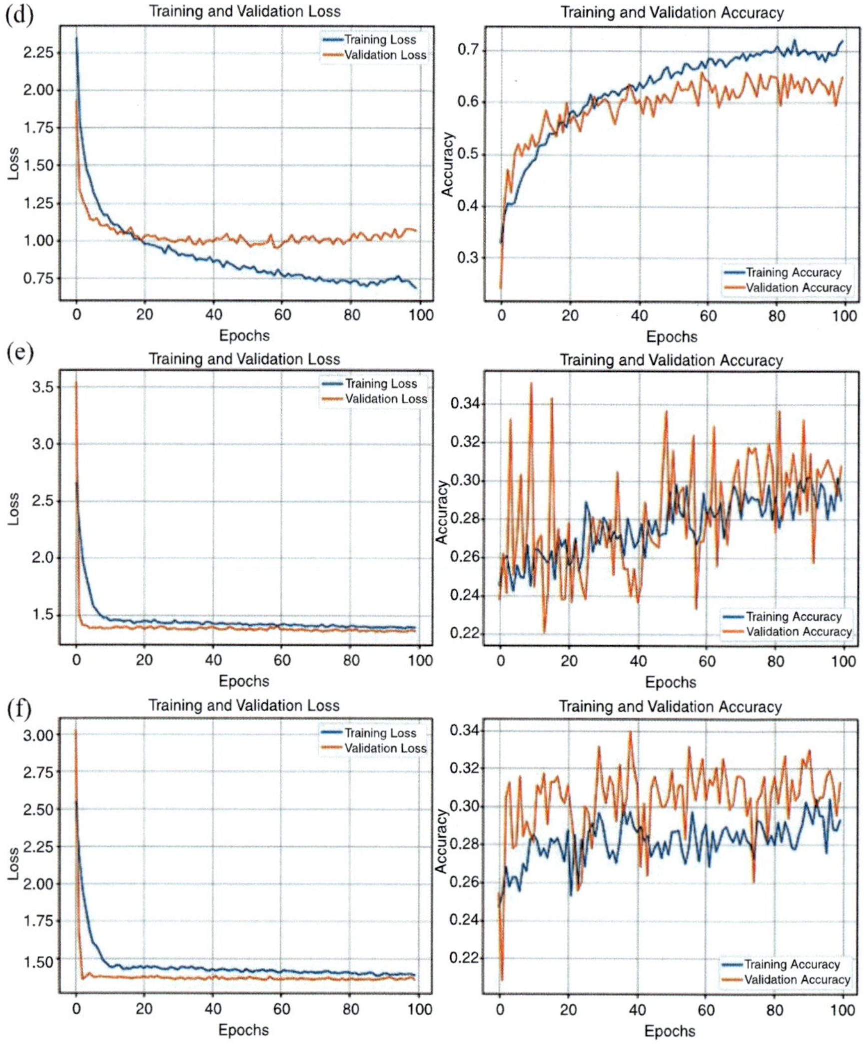

Fig. 4.4 (continued)

As could be observed from Fig. 4.4b, the training and validation loss of InceptionResNetV2 are decreasing to approximately 0.1 and 0.56, respectively, after 100 epochs. Regarding the accuracy curves, the training accuracy of the model is remarkably improved up to 82% from epoch 0 to 10, while 88% is achieved for the validation accuracy. As for convergence, the model converges after 87 epochs reaching a training accuracy of about 96%. A remark is also made that the validation loss starts to increase at epoch 10, and at the same time, the training loss is continuing to decline, which indicates an overfitting problem.

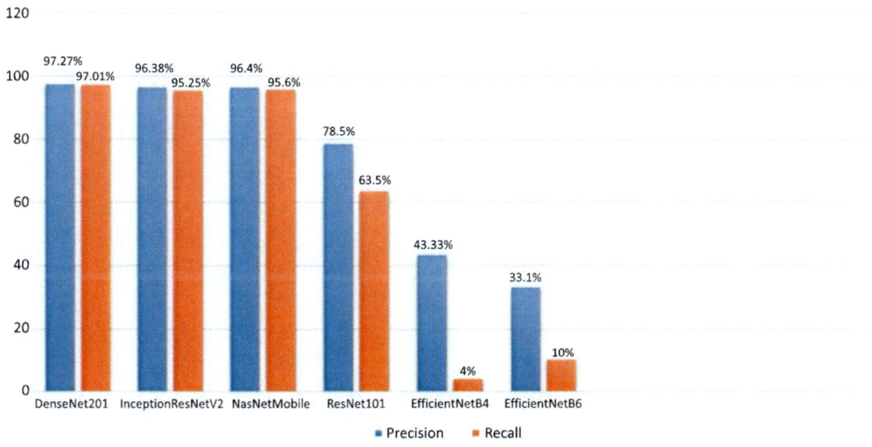

Fig. 4.5 Precision and recall evaluation of the used models

From the loss and accuracy curves of the NasNetMobile model, we can see that the training loss keeps decreasing and training accuracy keeps increasing until convergence. But the validation loss is decreasing, while the validation accuracy is still improving. Hence, it produces training accuracy of about 95% and validation accuracy of about 86%, respectively. Regarding the model loss, it achieves the training loss and the validation loss of about 0.1 and 0.56, respectively. However, at epoch 9, the model starts to overfit.

Concerning the loss curve of the ResNet101, the loss keeps decreasing and the model produces the training and validation loss of about 0.7 and 1.0, respectively, after 100 epochs, which means that the model does not generalize well on the validation samples. Correspondingly, if we perceive the accuracy curve of the ResNet101, the training accuracy gets a value close to 70%, while the validation accuracy reaches a value of about 65%.

In the loss plots of the EfficientNetB4 and EfficientNetB6 models, we can notice that both training and validation graphs are relatively stable compared to the previous-presented loss plots. The loss reached during the model training and validation is around 0.7. So, an underfitting problem occurs for both models. We also observe that both models generate the worst classification accuracy value of about 30%. As shown, among the six models developed, DenseNet201 yields the highest loss and accuracy for training and validation sets.

Figures 4.5, 4.6, and 4.7 report the performance comparison of the six CNN architectures in terms of classification precision, recall, sensitivity, specificity, and accuracy.

From the obtained results shown in Fig. 4.5, we clearly see that DenseNet201 delivers the highest psoriasis classification precision (97.27%) and recall (97.01%), followed by NasNetMobile with values of 96.4% and 95.6%, respectively. The InceptionResNetV2 and NasNetMobile also show considerable performance. The

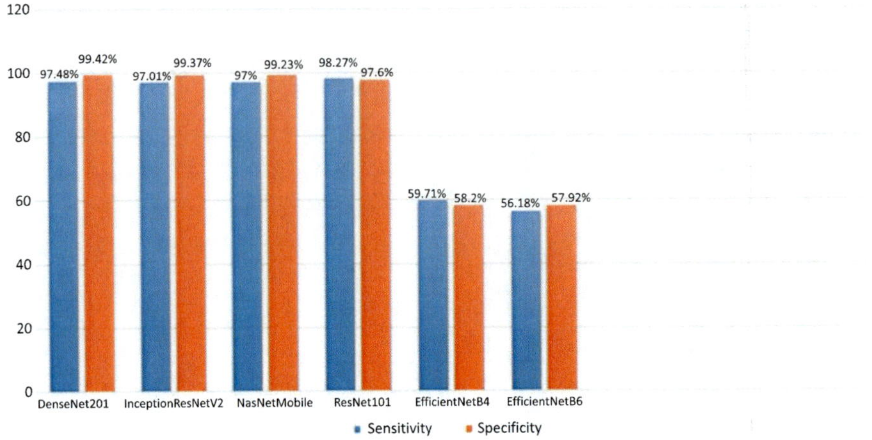

Fig. 4.6 Sensitivity and specificity evaluation of the used models

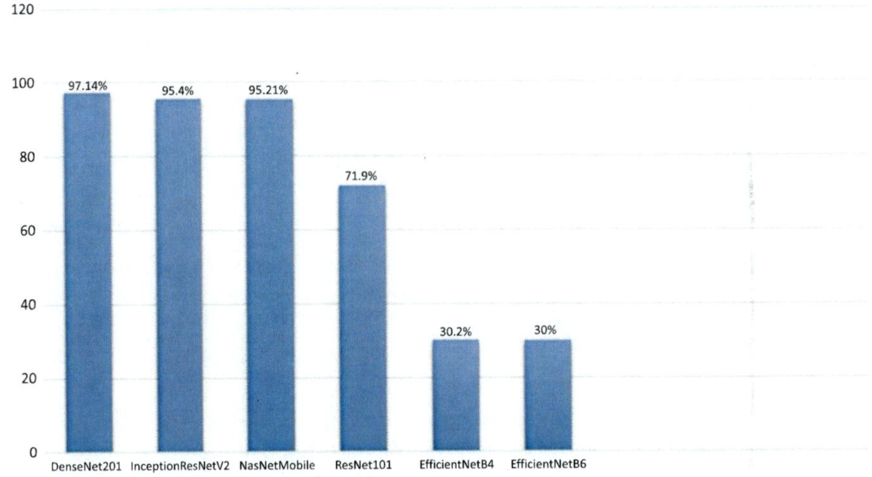

Fig. 4.7 Accuracy evaluation of the used models

first model contributes a precision (and a recall) of 96.38% (and 95.25%), while the second model achieves a precision of 96.4% and a recall of 95.6%. However, the EfficientNetB4 and EfficientNetB6 models produce on the whole the lowest precision and recall values.

As seen from Fig. 4.6, EfficientNetB4 and EfficientNetB6 provide the lowest classification sensitivity and specificity of around 50%, while the DenseNet201 produces the highest sensitivity of 97.48% and the highest specificity of 99.42%.

When comparing the DenseNet201 accuracy results to those of InceptionResNetV2, NasNetMobile, ResNet101, EfficientNetB4, and EfficientNetB6, it gives

Table 4.3 Training time evaluation of the used models

Classifier	Training time (per 100 epochs)
DenseNet201	50 minutes
InceptionResNetV2	80 minutes
NasNetMobile	30 minutes
Resnet101	35 minutes
EfficientNetB4	32 minutes
EfficientNetB6	90 minutes

better accuracy performance (97.14%) that proves its superiority to recognize the psoriasis skin lesions. This may be due to the effectiveness of the feature map generation method of the dense blocks.

Besides, Table 4.3 presents the training time of the used six models for the psoriasis classification problem.

It is observed from Table 4.3 that the NasNetMobile has the lowest training time of around 30 minutes per hundred epochs. DenseNet201 demands a moderate training time, of approximately 50 minutes, while InceptionResNetV2 and EfficientNetB6 require more training time (of about 80 and 90 minutes, respectively). These results could thus be justified by the depth of the designed architectures. The lower the network depth (i.e., lightweight architectures) is, the faster the training would be. Also, using a stack of CNN models (such as in InceptionResNetV2) for the classification yields in more training parameters and computational cost compared to using a single network.

4.6 Conclusion

In this chapter, we presented DenseNet201 architecture-based computer-aided diagnosis system for the psoriasis classification. The performance evaluation was performed on a database of 4212 images, belonging to the psoriasis, eczema, rosacea, and healthy skin images, respectively. The proposed psoriasis classification method based on DensNet201 model outperformed the InceptionResNetV2, NasNetMobile, ResNet101, EfficientNetB4, and EfficientNetB6 architectures. The achieved findings could be exploited in the assessment and the classification of the severity of psoriasis skin lesion images.

Finally, the current work has some limitations. First, the number of the skin lesion classes used to undertake the study was limited to four classes. Second, increasing the size of the dataset and the K-cross-validation sampling could improve the skin lesion diagnosis performances. Third, more transfer-learning-based CNN architectures should be applied, tested, and compared to solve the same psoriasis classification problem. Fourth, advanced machine learning algorithms (such as transformer) would potentially boost the outcomes quality.

References

1. S.F. Aijaz, S.J. Khan, F. Azim, C.S. Shakeel, U. Hassan, Deep learning application for effective classification of different types of psoriasis. J. Healthcare Eng. **2022**, 1–2 (2022)
2. K. Belattar, et al., A comparative study of CNN architectures for melanoma skin cancer classification (2022)
3. F. Chollet, Xception: deep learning with depthwise separable convolutions, in *Proceedings of the IEEE Conference on Computer Vision and Pattern Recognition* (2017), pp. 1251–1258
4. M. Dash, N.D. Londhe, S. Ghosh, R. Raj, R. Sonawane, Psoriasis lesion detection using hybrid seeker optimization-based image clustering. Curr. Med. Imaging **17**(11), 1330–1339 (2021)
5. A. Ghani, C.H. See, V. Sudhakaran, J. Ahmad, R. Abd-Alhameed, Accelerating retinal fundus image classification using artificial neural networks (ANNs) and reconfigurable hardware (FPGA). Electronics **8**(12), 1522 (2019)
6. K. He, X. Zhang, S. Ren, J. Sun, Deep residual learning for image recognition (2015). https://doi.org/10.48550/ARXIV.1512.03385, https://arxiv.org/abs/1512.03385
7. K. He, X. Zhang, S. Ren, J. Sun, Deep residual learning for image recognition, in *Proceedings of the IEEE Conference on Computer Vision and Pattern Recognition* (2016), pp. 770–778
8. A.G. Howard, M. Zhu, B. Chen, D. Kalenichenko, W. Wang, T. Weyand, M. Andreetto, H. Adam, MobileNets: efficient convolutional neural networks for mobile vision applications. Preprint. arXiv:1704.04861 (2017)
9. J. Hu, L. Shen, G. Sun, Squeeze-and-excitation networks, in *Proceedings of the IEEE Conference on Computer Vision and Pattern Recognition* (2018), pp. 7132–7141
10. G. Huang, Z. Liu, L. Van Der Maaten, K.Q. Weinberger, Densely connected convolutional networks, in *2017 IEEE Conference on Computer Vision and Pattern Recognition (CVPR)*, Honolulu, HI, USA, pp. 2261–2269. https://doi.org/10.1109/CVPR.2017.243
11. G. Huang, Z. Liu, L. Van Der Maaten, K.Q. Weinberger, Densely connected convolutional networks, in *Proceedings of the IEEE Conference on Computer Vision and Pattern Recognition* (2017), pp. 4700–4708
12. G. Kalkan, Comorbidities in psoriasis: the recognition of psoriasis as a systemic disease and current management. Turkderm-Archives of the Turkish Dermatol. Venereol. **51**(3), 1–7 (2017)
13. G.W. Kimmel, M. Lebwohl, Psoriasis: overview and diagnosis. Evidence-Based Psoriasis. Hindawi J. Healthcare Eng., **2022**, 7541583, 12 pages, (2018). https://doi.org/10.1155/2022/7541583
14. G.W. Kimmel, M. Lebwohl, Psoriasis: Overview and Diagnosis. In: Bhutani, T., Liao, W., Nakamura, M. (eds) Evidence-Based Psoriasis. Updates in Clinical Dermatology. (Springer, Cham. 2018). https://doi.org/10.1007/978-3-319-90107-7_1
15. G. Krueger, J. Koo, M. Lebwohl, A. Menter, R.S. Stern, T. Rolstad, The impact of psoriasis on quality of life: results of a 1998 National Psoriasis Foundation patient-membership survey. Arch. Dermatol. **137**(3), 280–284 (2001)
16. L.F. Li, X. Wang, W.J. Hu, N.N. Xiong, Y.X. Du, B.S. Li, Deep learning in skin disease image recognition: a review. IEEE Access **8**, 208264–208280 (2020)
17. H. Li, Y. Pan, J. Zhao, L. Zhang, Skin disease diagnosis with deep learning: a review. Neurocomputing **464**, 364–393 (2021)
18. G. Mahrle, H.J. Schulze, L. Färber, G. Weidinger, G.K. Steigleder, et al.: Low-dose short-term cyclosporine versus etretinate in psoriasis: improvement of skin, nail, and joint involvement. J. Am. Acad. Dermatol. **32**(1), 78–88 (1995)
19. A. Mikołajczyk, M. Grochowski, Data augmentation for improving deep learning in image classification problem, in *2018 International Interdisciplinary PhD Workshop (IIPhDW)* (IEEE, Piscataway, 2018), pp. 117–122
20. S.J. Pan, Q. Yang, A survey on transfer learning. IEEE Trans. Knowl. Data Eng. **22**(10), 1345–1359 (2009)
21. L. Peng, Y. Na, D. Changsong, L. Sheng, M. Hui, Research on classification diagnosis model of psoriasis based on deep residual network. Digit. Chin. Med. **4**(2), 92–101 (2021)

22. Z. Rahman, M.S. Hossain, M.R. Islam, M.M. Hasan, R.A. Hridhee, An approach for multiclass skin lesion classification based on ensemble learning. Inf. Med. Unlocked **25**, 100659 (2021)
23. K. Simonyan, A. Zisserman, Very deep convolutional networks for large-scale image recognition. Preprint. arXiv:1409.1556 (2014)
24. C. Szegedy, W. Liu, Y. Jia, P. Sermanet, S. Reed, D. Anguelov, D. Erhan, V. Vanhoucke, A. Rabinovich, Going deeper with convolutions, in *Proceedings of the IEEE Conference on Computer Vision and Pattern Recognition* (2015), pp. 1–9
25. C. Szegedy, S. Ioffe, V. Vanhoucke, A.A. Alemi, Inception-v4, Inception-ResNet and the impact of residual connections on learning, in *Thirty-First AAAI Conference on Artificial Intelligence* (2017)
26. M. Tan, Q.V. Le, EfficientNet: rethinking model scaling for convolutional neural networks. CoRR abs/1905.11946 (2019), http://arxiv.org/abs/1905.11946
27. B. Tuzun, The differential diagnosis of psoriasis vulgaris. Pigmentary Disord. **3**(245), 2376–0427 (2016)
28. S. Umapathy, M. Sampath, S. Srivastava, et al., Automated segmentation and classification of psoriasis hand thermal images using machine learning algorithm, in *Proceedings of the International e-Conference on Intelligent Systems and Signal Processing* (Springer, Berlin, 2022), pp. 487–496
29. S. Xie, R. Girshick, P. Dollár, Z. Tu, K. He, Aggregated residual transformations for deep neural networks, in *Proceedings of the IEEE Conference on Computer Vision and Pattern Recognition* (2017), pp. 1492–1500
30. Y. Yang, J. Wang, F. Xie, J. Liu, C. Shu, Y. Wang, Y. Zheng, H. Zhang, A convolutional neural network trained with dermoscopic images of psoriasis performed on par with 230 dermatologists. Comput. Biol. Med. **139**, 104924 (2021)
31. X. Zhang, S. Wang, J. Liu, C. Tao, Towards improving diagnosis of skin diseases by combining deep neural network and human knowledge. BMC Med. Inf. Decision Making **18**(2), 69–76 (2018)
32. B. Zoph, V. Vasudevan, J. Shlens, Q.V. Le, Learning transferable architectures for scalable image recognition, in *Proceedings of the IEEE Conference on Computer Vision and Pattern Recognition* (2018), pp. 8697–8710

Part II
New Perspectives in Computational Intelligence

Chapter 5
A Novel Embryonic Cellular Architecture with BIST for Deep Space Systems

Gayatri Malhotra, Punithavathi Duraiswamy, and J. K. Kishore

5.1 Introduction

Normally in the ground systems and near-earth space systems, the fault tolerance approaches used are based on redundancy at system level and triple modular redundancy (TMR) at FPGA circuit level. In contrast to this, the deep space systems require a different approach for fault tolerance. The deep space systems need to be resource efficient and power efficient. These systems should have ability to re-configure itself to counter new challenges. The present trend of micro- and nano-satellites for deep space missions has called for a different fault tolerance technique. As the size of satellite is reduced, the resources are also limited. The embryonic fabric-based cellular architecture provides self-repair using additional spare cells instead of triplicating all the cells. Thus this approach is suitable for smaller space system design in deep space missions.

The field of bio-inspired systems has emerged significantly during last few decades. The "Evolvable," "Embryonics," and "Self-repair" are the few terms mostly cited. To design the electronics system fault-tolerant, the embryonic design approach is considered [1, 2]. The Actel and Virtex FPGA contains logic cells and configurable logic blocks (CLB), connected through reconfigurable interconnects.

G. Malhotra (✉)
U R Rao Satellite Centre, Bangalore, India

M S Ramaiah University of Applied Sciences, Bangalore, India
e-mail: gayatri@ursc.gov.in

P. Duraiswamy
M S Ramaiah University of Applied Sciences, Bangalore, India

J. K. Kishore
U R Rao Satellite Centre, Bangalore, India

M. D. Hina et al. (eds.), *Future Research Directions in Computational Intelligence*, EAI/Springer Innovations in Communication and Computing,
https://doi.org/10.1007/978-3-031-34459-6_5

In Virtex FPGA CLB, it contains LUT, flip-flops, adder, MUX, and boundary scan components. Equivalent to FPGA architecture, in the proposed embryonic fabric, there is multi-cellular structure, where each cell contains logical block, flip-flops, BIST control, and additional self-repair unit.

In present space system design, the triple modular redundancy (TMR) approach is used in FPGA to isolate single-point failure and utilizes high resources of FPGA to implement it, while, in the proposed embryonic fabric, additional spare cells are to be utilized in case of faulty cell detection. The CGP format represents the digital circuit as a rectangular array of nodes. Each node is an operation on the node inputs. All node inputs, node operations, and node outputs are sequentially indexed by integers. The configuration data (Genome data) for the embryonic cell are a linear string of these integers.

Applying evolutionary algorithm (EA) for the digital circuit design explores the larger search space. The design optimization is better achieved by the EA than the traditional approaches [3, 4]. In this chapter, a novel embryonic fabric architecture with in-built self-test module is designed and simulated using verilog. The configuration genome data generation is through EA, while the genome data format is Cartesian Genetic Programming (CGP) [5, 6]. The CGP data generation is through HsClone algorithm, while other EA are also planned to be tested [7].

5.1.1 Embryonics: Emerging Trend in Deep Space Systems

Embryonics being a homogeneous array of embryonic cells possesses self-healing due to its structure. The embryonics is also known as "Electronic Stem Cells" [8]. It implements digital system with fault-tolerant capabilities inspired from ontogeny process. The embryonic cellular structure is capable to induce fault tolerance in the design by self-repairing [9, 10]. The natural processes of growing, reproducing, and healing are adopted in the electronic design. The "Stem Cell" has feature of becoming anything they want to be, and this has inspired the electronics design using embryonic cellular structure. The mechanism of division is the main driving force behind the development of entire organism. The same is applied as cloning mechanism in this embryonic fabric design.

There are different approaches to implement self-healing, self-repair, and self-replicate in the embryonic cellular architecture. The electronic tissue called POEtic design [11] is inspired from three life axes of phylogenesis, ontogenesis, and epigenesis. The ontogenetic axis refers to the cellular growth that helps in self-repair. The structural principles of living organism such as multi-cellular architecture, cellular division, and cellular differentiation are utilized to enable systems to grow, self-replicate, and self-repair [12]. The embryonic cell architecture is to be designed for self-diagnostic along with the fault recovery methods embedded in it. The approach described in [13] is to kill the faulty cell and make it transparent, while transferring its functions to neighboring cell. The self-healing is achieved through cell elimination and further by row elimination. In eDNA approach [14], the electronic cell "eCell" reads the electronic DNA "eDNA" to interpret the function it has to do and also in case of one eCell failure, to move the function to another eCell. Thus, to create self-organization and self-healing of electronic cells.

In this chapter, a built-in self-test (BIST) methodology [15] is integrated with in the fabric to initiate the self-repair at embryonic cell level. The BIST controller controls the different types of embryonic cell as per their selected gene. The fault detection information at embryonic cell level is available to embryonic fabric controller. Based on this, the controller initiates the self-repair of the faulty cell. This approach deals with transient errors and provides self-healing [16].

In Sect. 5.2, the novel embryonic fabric design for combinatorial digital circuits is proposed. It describes the design of embryonic cell and switch box interconnection. In Sect. 5.3, the built-in self-test design methodology for embryonic cells is presented. The sub-modules of the design such as controller, random pattern generator (RPG), response analyzer, and fault detection method are underlined. Section 5.4 is about fault detection results for adder and comparator cells. Section 5.5 contains the simulation results. Section 5.6 concludes the results, and Sect. 5.7 is the scope for future work.

5.2 Embryonic Fabric for Digital Circuits

The proposed embryonic fabric consists of embryonic cells, switch boxes, and the fabric controller module. The fabric controller has to control the data transfer between the cells and switch boxes along with input–output units. The embryonic fabric architecture is shown in Fig. 5.1. The embryonic fabric operation can be defined in the following steps:

1. Fabric controller initiates the Genome data transfer to first embryonic cell's memory.
2. Genome data get copied to all the cells through cloning process.
3. CGP decoder in each cell decodes the data based on gene-sel information, and cell functionality is defined.
4. For modular design, required signals are transferred through switch boxes.
5. fpgain is fed to each cell, and fpgaout is available after cell function.
6. Fabric controller initiates BIST controller for each cell's function.
7. On fault detection, identified cell's self-repair module becomes active.
8. Self-repair initiates scrubbing of faulty cell through its own memory data: for transient faults.

In the FPGA structure, a fixed amount of configuration bits are required to configure its logic blocks and to establish interconnections. In the embryonic cell fabric, it is achieved with less configuration bits as data generation is through EA. The cloning process in the fabric implements the data copy to all the cells. The first cell's genome data are loaded externally, and cloning takes care of copying the genome data bits to next cells during runtime. The cloning process is discussed earlier where data

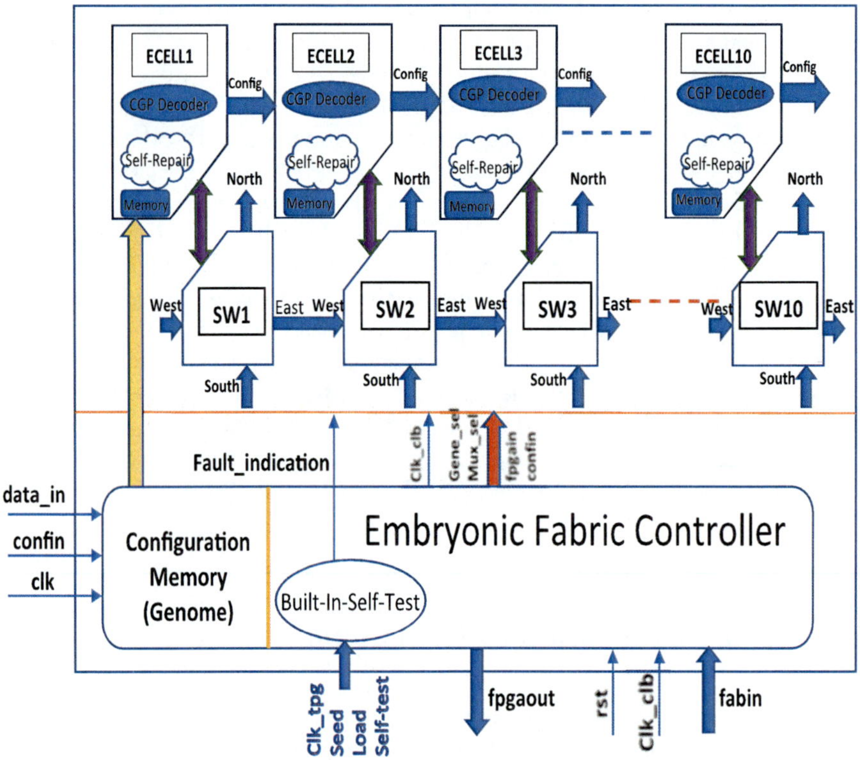

Fig. 5.1 Embryonic fabric architecture

are in look-up table (LUT) format [17]. The circuit size is flexible and contains clone data bits along with CGP data. The genome data contain two genes, adder, and comparator functions. The gene data are decoded by CGP decoder within the embryonic cell. The genome data are 161 bits that contain CGP data for 1-bit adder and 2-bit comparator. The fabric controller integrates BIST controller to enable self-test. The self-test is established at cell level to control self-repair also at cell level. The proposed embryonic fabric is tested for combinational and sequential circuits, so making it suitable for digital design of space systems.

5.2.1 Embryonic Adder and Comparator Cell Design

The embryonic cell contains genome data memory, cloning mechanism, CGP decoder, cell configuration controller, and self-repair module. The genome data memory stores the data in CGP format. The cloning of genome data into next cells is performed till the "total-clone-cnt" is greater than zero. The "confin" is the trigger

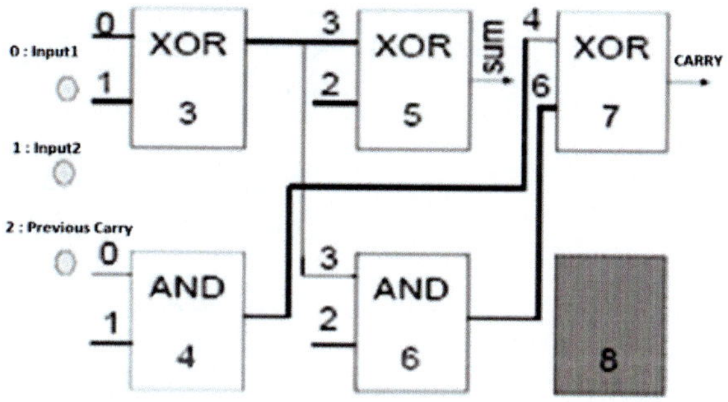

Fig. 5.2 CGP configuration of 1-bit adder

signal to initiate genome data loading into first cell. The data memory can be loaded by self-repair module when there is fault detection in the cell. The cascading of cells is possible to have scalable circuit design. The 4-bit adder is designed from 1-bit adder configuration data, and 8-bit comparator is designed from 2-bit comparator configuration data. The embryonic cell integrates the CGP decoder that decodes configuration data for adder and comparator cells.

The "fpgain" is the external input applied to CGP decoder. The switch box is used to route the signals between the cells to implement cascading. The adder cells transfer carry bits through switch box. The comparator cells transfer lower bit-level outputs Iy $(A < B)$ and Iz $(A > B)$ to higher bit-level comparator. In CGP, the representation of data is in the form of nodes. Each node is expressed as $in1, in2, logicalfunction$. The K-map derived data for 1-bit adder are 0 1 0; 0 1 1; 3 2 0; 3 2 1; 6 4 2; 5; 7. The K-map derived data for 2-bit comparator are 132; 13f; 31f; 02b; 048; 47c; 28b; 699; a59; b; c. The CGP configuration of adder is shown in Fig. 5.2 and of comparator is shown in Fig. 5.3. The CGP data length for adder and comparator data is 161 bits (45 + 4 bits:adder + 108 + 4 bits:comparator). The adder has five nodes (5×3 octets $\times 3$ bits = 45 bits), and comparator has nine nodes (9×3 octets $\times 4$ bits = 108 bits). The clone count for each adder or comparator cell is expressed using four bits within configuration data. The evolutionary algorithm is integrated with the fabric design to generate optimized CGP data (genome). The details of EA are not part of this paper.

5.2.2 *Embryonic Switch Box Design*

The data transfer between the cells are done through switch boxes. Each switch box has four directional buses and one dedicated bus with the neighboring cell. The adder and comparator functions are carried out using east and west data buses. Buses are used to route "carry" signal for adder cells and "AsmlB," "AlargB" signals for comparator cells. The signals between embryonic cell and switch box are used in CGP data decoding.

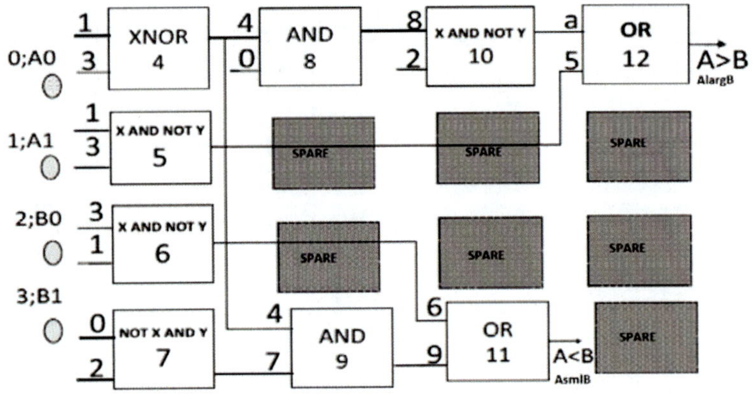

Fig. 5.3 CGP configuration of 2-bit comparator

5.3 Built-in Self-test Design for Embryonic Cells

BIST is a design for testability (DFT) that includes the testing features within the circuit under test (CUT). The basic BIST architecture requires the following modules within the embryonic fabric design controller:

1. A test controller
2. A test pattern generator
3. A response analyzer

A typical BIST architecture is shown in Fig. 5.4. The test pattern generator (TPG) is to generate the test patterns for CUT. The selected type of TPG is linear feedback shift register (LFSR). A response analyzer (RA) is required to compare the CUT response with stored response. It is designed using LFSR that is used as a signature analyzer. It compacts and analyzes the CUT test responses to determine any fault. A test controller block activates the self-test mode and analyzes the responses. As shown in Fig. 5.4, the primary inputs to MUX and CUT outputs to primary outputs (PO) cannot be tested by BIST. During normal operation mode, CUT (embryonic fabric) receives its inputs from test bench/simulator and performs the function for which it is designed. During self-test mode, a TPG circuit applies a sequence of test patterns to the CUT. The test responses are evaluated by response compactor. The responses are compacted to form *signatures*. The reference golden *signatures* are already saved for no fault condition. The response *signature* is compared with the stored reference signature to find if CUT is good or faulty.

The following are the design parameters that are to be considered while developing BIST methodology:

- Fault coverage: The fraction of faults that can be detected by response monitor. In case of the presence of input bit stream errors, the computed signature can match with golden signature. This is called *aliasing*.

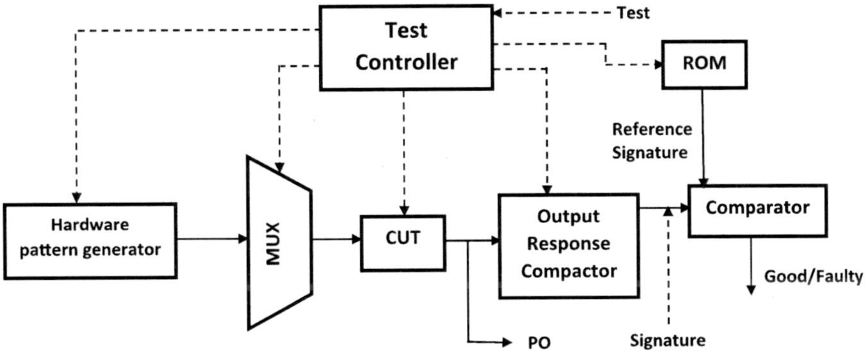

Fig. 5.4 A typical BIST architecture

- Test set size: Large test sets imply better fault coverage.
- Hardware overhead: Extra hardware is required to implement BIST function.
- Performance overhead: The impact of BIST hardware to path delays to affect normal operation.

5.3.1 BIST Controller Design for Embryonic Fabric

The BIST controller as a part of embryonic fabric controller initiates self-test mode whenever triggered. The "fpgain" are inputs generated through test bench as "fabin" in normal mode and as "testin" in self-test mode. In case the self-test is planned, the BIST controller initiates test pattern generator module and response analysis module. The clock for BIST modules is "clk-tpg," simulated in self-test mode only. The "fpgaout" from the embryonic cell corresponding to the "testin" is routed to the controller to response analyzer. The fault indication from the RA is transferred to faulty cell for further action. The faulty cell memory scrubbing is attempted as part of self-repair action.

5.3.2 Test Pattern Generator Module

The TPG is implemented using LFSR, a 40-stage shift register formed from flip-flops (FFs), with the outputs of selected FFs being fed back to the shift register's inputs. When LFSR is used for TPG, it has to cycle rapidly through a large number of states. These states are defined by the design parameters of the LFSR and generate the test patterns. In this work, the BIST TPG generates the pseudo-random binary sequence. The pseudo-random pattern BIST needs lower hardware

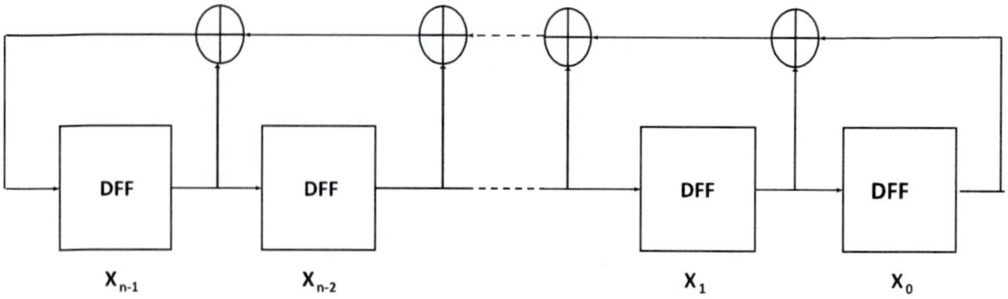

Fig. 5.5 A typical n-stage linear feedback shift register

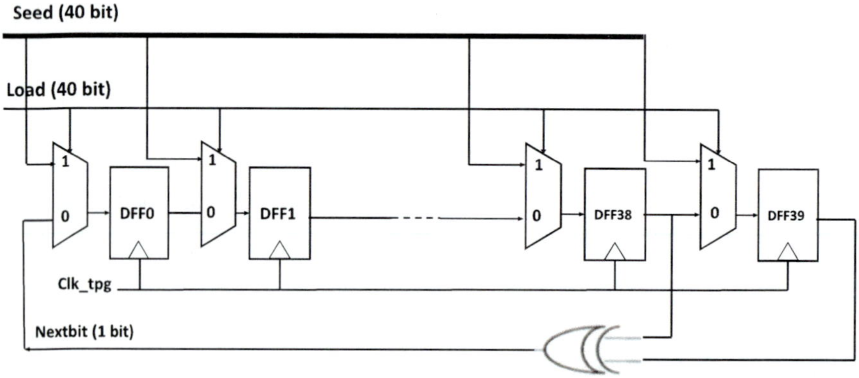

Fig. 5.6 Test pattern generator 40-bit LFSR

and design efforts than other methods such as exhaustive test. The linear feedback shift register reseeding-based BIST technique is used to control the LFSR state. The output patterns of the LFSR are time-shifted and become correlated. Therefore, a network of XOR gates are added to decorrelate the output patterns of LFSR. A typical n-stage LFSR with external XOR is shown in Fig. 5.5. The implemented LFSR is 40-stage with the characteristic polynomial equation as

$$P(X) = X^{39} + X^{38} + 1. \tag{5.1}$$

The final LFSR configuration to generate 40-bit test data is shown in Fig. 5.6. LFSR is initially loaded with seed data when load is "1." The LFSR function is synchronous with "clk-tpg."

5.3.3 *Output Response Analysis Module*

The RA module has to check the CUT responses for the random test pattern applied as input. In the embryonic fabric of ten cells, the response from all the cells is to be verified. Each adder embryonic cell has two outputs—sum and carry. Each comparator embryonic cell also has two outputs—AsmlB and AlargB. As per gene selected for a cell, the function is identified as adder or comparator. The total response is enormous, so it is required to compact this response to manageable size. The RA compresses a very large test response into a single word called as *signature*. The *signature* is compared with the pre-stored *golden signature* obtained from the fault-free circuit responses using the same compression mechanism. If the *signature* matches the *golden signature*, the CUT is fault-free. Otherwise, it is faulty.

The response analyzer technique used for embryonic fabric is *signature analysis*. The response compaction is done using LFSR. The data bits from POs are compacted by dividing the PO polynomial by characteristic polynomial of LFSR. The remainder of the polynomial division is the *signature*. The seed value is usually zero before testing.

MISR for Response Compaction—Signature Analysis

In ordinary LFSR response compacter, one of these has to put for each PO, and this will lead to hardware overhead. Multiple-input signature register (MISR) is the type that compacts all the cell outputs into one LFSR. All responses of adder cells are superimposed into one "signature-reg-adder." Similarly, all responses of comparator cells are superimposed into "signature-reg-cmprtr." The design of adder MISR and comparator MISR is shown in Figs. 5.7 and 5.8.

Both MISRs are 10-stage LFSR as there are five adder cells (four cascaded and one spare) with two outputs from each and five comparator cells with two outputs from each. The implemented LFSR is internal feedback type and has characteristic polynomial as

$$P(X) = X^9 + X^8 + 1. \tag{5.2}$$

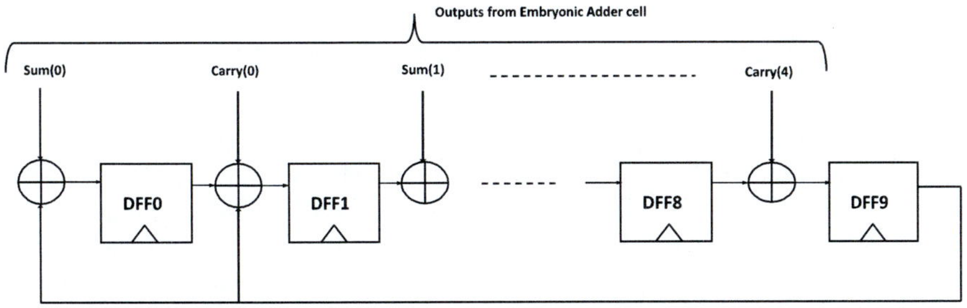

Fig. 5.7 MISR for embryonic adder cells outputs

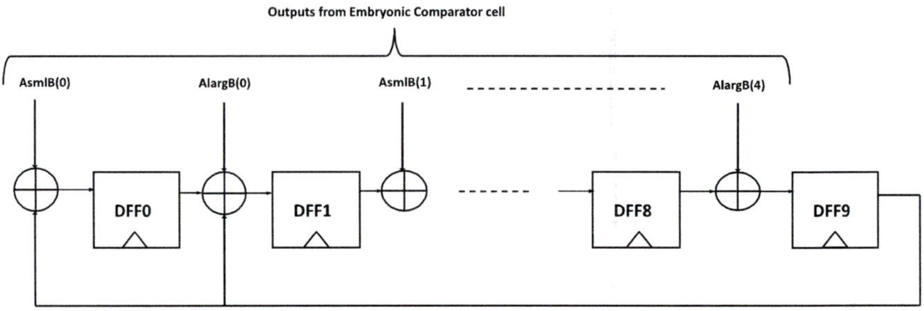

Fig. 5.8 MISR for embryonic comparator cells outputs

5.4 Fault Detection of Embryonic Adder and Comparator Cells

For fault detection, the degree of data polynomial (output response from cells) should be less than $2^{10} - 1$, where 10 is the degree of the characteristic polynomial of the LFSR. The number of "clk-tpg" cycles are calculated (about 1000), and after that, fault-free signature analysis register (SAR) is saved. After the SAR is saved for adder and comparator, the MISR gets reset and loaded with zero value. The TPG module is loaded with the same seed as was at the start of self-test. After the same clock cycles, self-test signal is made low, and the "signature-reg-adder" and "signature-reg-cmprtr" are saved. The "signature" value for adder and comparator is compared with the saved fault-free signature value. The fault is simulated by making one of the cell outputs as "stuck at 0." This is successfully detected by the RA module. The fault detection signal is routed to fabric controller to faulty cells. As fault is identified, self-repair is initiated by re-loading the configuration data to all faulty cells (scrubbing). The reliability analysis of self-repair is to be included [18].

5.5 Fault Simulation Results of Embryonic Cell

The embryonic fabric with ten cells is created through configuration data cloning. The cloning process is shown in Fig. 5.9. The genome data are copied to ten cells once "confin" is high. The self-test mode of the embryonic fabric is triggered with "self-test" signal. The simulation is shown in Fig. 5.10. The "self-test" is set high, and the "clk-tpg" is also initiated with it. The "fpgain" is the input simulation from test bench, while the "testin" is the input simulation from BIST module. Once the "self-test" is high, the "seed" is loaded into LFSR for random test pattern generation. The "testin" is applied to embryonic fabric cells as inputs, the corresponding "fpgaout," the outputs from cells are saved. The "gene-sel" for ten cells is defined as

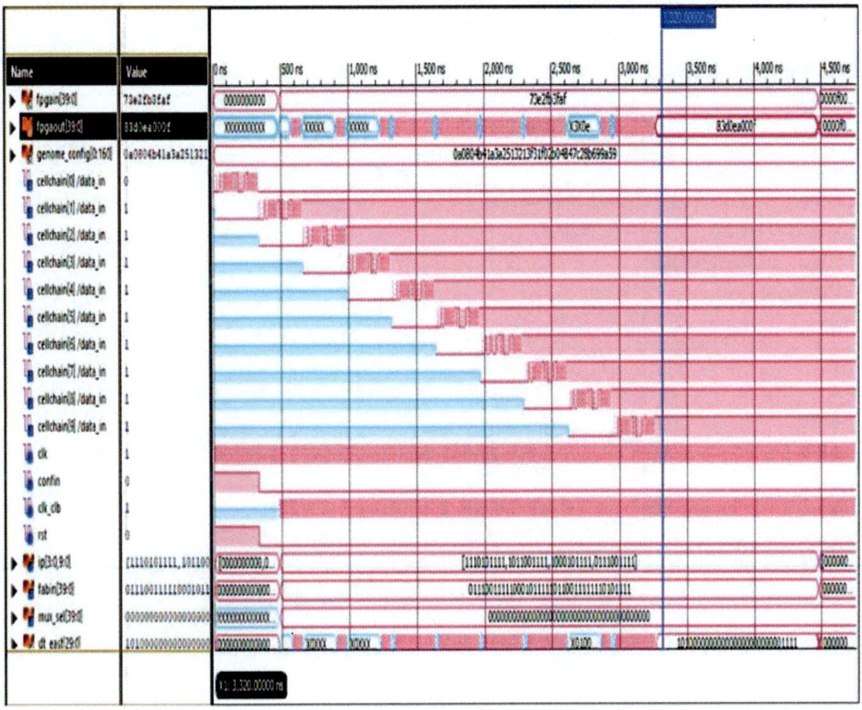

Fig. 5.9 Embryonic fabric cell cloning mechanism

"0" for adder cell and "1" for comparator cell. The switch box buses "dt-east" and "dt-west" are assigned with the signals that need to be transferred between cells. The "fault indication" is the one-bit signal from each cell. After the fixed clock cycles, the *signatures* are saved.

The RA module signals are depicted in Figs. 5.11 and 5.12. The "sig-save" is triggered after fixed clocks to load the signature registers of adder and comparator MISR. The "load" signal is high initially for test pattern generation and again set high when "sig-save" is high. This will reset the TPG-LFSR with seed value, and again the fixed clock cycles are executed. In Fig. 5.12, after "signature-reg-adder" and "signature-reg-cmprtr" are saved, then after fixed clock cycles, "fault indication" is analyzed by comparing "state-out-adder" with "signature-reg-adder" and "state-out-cmprtr" with "signature-reg-cmprtr." At the time of fault detection, "self-test" signal is set low to stop random pattern generation.

The sum output from all adder cells is simulated for "stuck at '0,'" and thus fault indication is high ("0100001111") for adder cells only. This indication is routed through fabric controller to each cell. The cell initiates "scrubbing" by re-loading genome data to cell memory. This is the self-repair process shown in Fig. 5.13. The "confin-test" initiates the faulty cell's scrubbing. The scrubbing is shown as

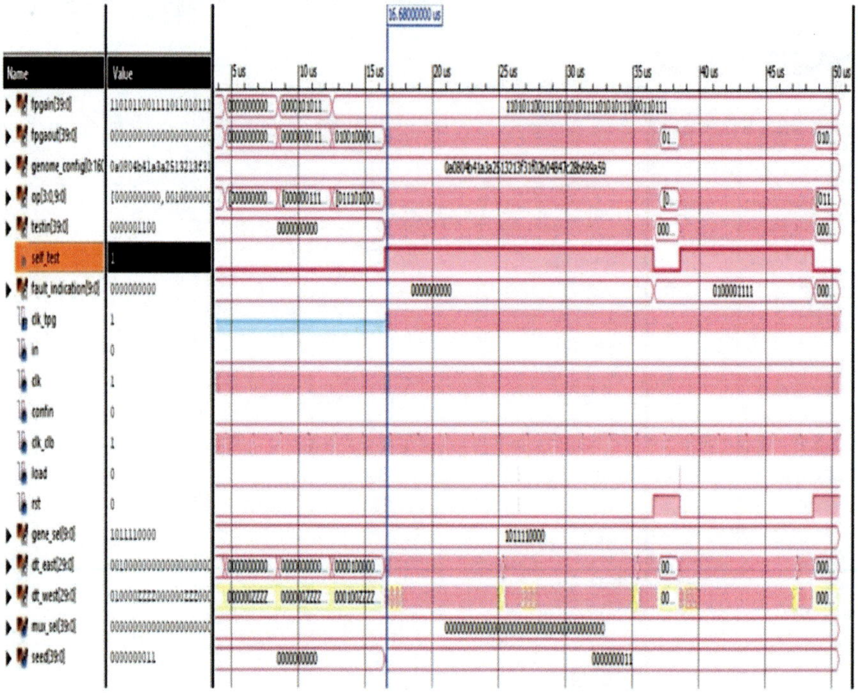

Fig. 5.10 Simulation results when self-test is triggered

serial data loading in Fig. 5.13. After it, "scrub-complete" is set high, to indicate the completion of process. To verify the fabric functionality, the RA is executed for same fixed clock cycles. As "signatures" are saved earlier, only comparison with MISR "state-out" is carried out. To check it, simulation for "stuck at '0'" is removed now. Thus, the simulation results indicate for no "fault indication." This approach is applicable for transient errors, while for the permanent error, the cell replacement is to be considered.

5.6 Conclusion

The novel embryonic fabric architecture designed with two types of genes, adder and comparator, is proposed for deep space systems. The BIST is implemented for the novel embryonic fabric architecture. The transient error is simulated for "stuck at zero" in adder cells. The fault detection is carried out and memory scrubbing got executed. After the memory scrubbing, the BIST module is re-initiated to verify the effect. The LFSR for random number generator is designed and run for fixed

Fig. 5.11 Simulation results of response analyzer:start

clock cycles. The response analyzer is implemented using MISR and designed separately for adder and comparator cells. The simulation results are verified for the implementation of self-repair through scrubbing.

5.7 Scope for Future Work

The fabric design has to be modified to include sequential circuits also. The BIST approach has to be upgraded in case of permanent error in the cells. In the case of permanent error, the faulty cell has to become transparent. The signal routing will be updated to use spare cell in place of faulty cell. The CGP data generation through other evolutionary algorithms also needs to be tested. The reliability analysis for the self-repairing strategy is to be carried out.

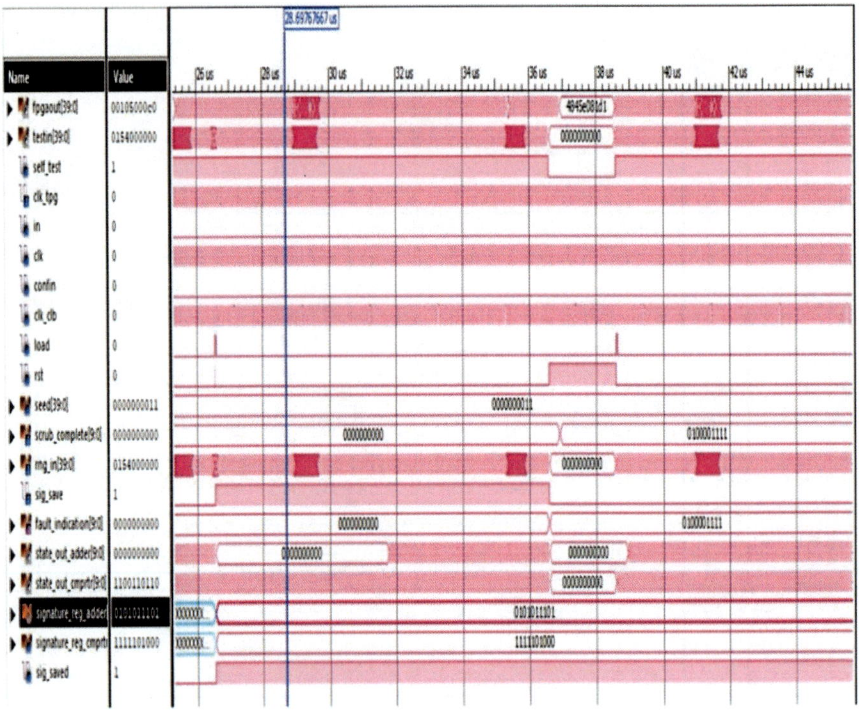

Fig. 5.12 Simulation results of response analyzer:end

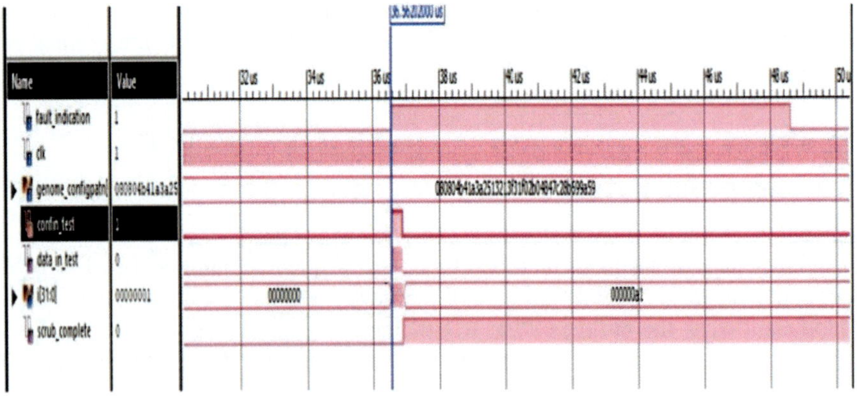

Fig. 5.13 Self-repair process through scrubbing

References

1. E. Benkhelifa, A. Pipe, A. Tiwari, Evolvable embryonics: 2-in-1 approach to self-healing systems. Procedia CIRP **11**, 394–399 (2013). https://doi.org/10.1016/j.procir.2013.07.029
2. V. Sahni, V.P. Pyara, An embryonic approach to reliable digital instrumentation based on evolvable hardware. IEEE Trans. Instrum. Meas. **52**(6), 1696–1702 (2003). https://doi.org/10.1109/TIM.2003.818737
3. K.H. Chong, I.B. Aris, M.A. Sinan, B.M. Hamiruce, Digital circuit structure design via evolutionary algorithm method. J. Appl. Sci. **7**, 380–385 (2007)
4. E. Benkhelifa, A. Pipe, G. Dragffy, M. Nibouche, Towards evolving fault tolerant biologically inspired hardware using evolutionary algorithms, in *2007 IEEE Congress on Evolutionary Computation*, Singapore (2007), pp. 1548–1554. https://doi.org/10.1109/CEC.2007.4424657
5. J.F. Miller, Cartesian genetic programming. Nat. Comput. Ser. **43**(June) (2011). https://doi.org/10.1007/978-3-642-17310-3
6. G. Malhotra, V. Lekshmi, S. Sudhakar, S. Udupa, Implementation of threshold comparator using cartesian genetic programming on embryonic fabric. Adv. Intell. Syst. Comput. **939**, 93–102 (2019)
7. E. Stomeo, T. Kalganova, C. Lambert, A novel genetic algorithm for evolvable hardware, in *2006 IEEE Congress on Evolutionary Computation*, CEC 2006, May (2006), pp. 134–141. https://doi.org/10.1109/CEC.2006.1688300
8. L. Prodan, G. Tempesti, D. Mange, A. Stauffer, Embryonics: electronic stem cells, in *Proceedings of the Eighth International Conference on Artificial Life (ICAL 2003)* (MIT Press, Cambridge, 2002), pp. 101–105
9. Y. Shanshan, W. Youren, A new self-repairing digital circuit based on embryonic cellular array, in *Proceedings of the ICSICT-2006: 2006 8th International Conference on Solid-State and Integrated Circuit Technology* (2006), pp. 1997–1999. https://doi.org/10.1109/ICSICT.2006.306573
10. D. Mange, A. Stauffer, G. Tempesti, Embryonics: a macroscopic view of the cellular architecture, in *Lecture Notes in Computer Science (Including Subseries Lecture Notes in Artificial Intelligence and Lecture Notes in Bioinformatics)*, vol. 1478 (1998), pp. 174–184. https://doi.org/10.1007/BFb0057619
11. Y. Thoma, G. Tempesti, E. Sanchez, POEtic: an electronic tissue for bio-inspired cellular applications. Biosystems **76**, 1–3 (2004)
12. A. Stauffer, D. Mange, J. Rossier, Design of self-organizing bio-inspired systems, in *Second NASA/ESA Conference on Adaptive Hardware and Systems (AHS 2007)* (2007)
13. X. Zhang, G. Dragffy, A.G. Pipe, N. Gunton, Q.M. Zhu, A reconfigurable self-healing embryonic cell architecture, in *Proceedings of the International Conference on Engineering of Reconfigurable Systems and Algorithms* (2003), pp. 134–140
14. M.R. Boesen, J. Madsen, eDNA: a bio-inspired reconfigurable hardware cell architecture supporting self-organisation and self-healing, in *Proceedings – 2009 NASA/ESA Conference on Adaptive Hardware and Systems, AHS 2009* (2009), pp. 147–154. https://doi.org/10.1109/AHS.2009.22
15. C.E. Stroud, *A Designer's Guide to Built-in Self-Test* (Springer, Berlin, 2002)
16. R. Salvador, A. Otero, J. Mora, E. de la Torre, L. Sekanina, T. Riesgo, Fault tolerance analysis and self-healing strategy of autonomous, evolvable hardware systems, in *Proceedings – 2011 International Conference on Reconfigurable Computing and FPGAs, ReConFig 2011* (2011), pp. 164–169. https://doi.org/10.1109/ReConFig.2011.37
17. G. Malhotra, J. Becker, M. Ortmanns, Novel field programmable embryonic cell for adder and multiplier, in *9th Conference on Ph.D. Research in Microelectronics and Electronics (PRIME-2013)* (2013)
18. Z. Zhang, Y. Wang, Method to self-repairing reconfiguration strategy selection of embryonic cellular array on reliability analysis, in *Proceedings of the 2014 NASA/ESA Conference on Adaptive Hardware and Systems, AHS 2014* (2014), pp. 225–232. https://doi.org/10.1109/AHS.2014.6880181

Chapter 6
Hybrid Whale Optimization Algorithm with Simulated Annealing for the UAV Placement Problem

Check for updates

Sylia Mekhmoukh Taleb, Yassine Meraihi, Selma Yahia, Amar Ramdane-Cherif, Asma Benmessaoud Gabis, and Dalila Acheli

6.1 Introduction

Unmanned aerial vehicles (UAVs) have drawn a lot of interest as potential options for wireless communication. UAVs were used in a large variety of applications such as security and surveillance, forest fire monitoring, agriculture, radiation and monitoring, meteorology, etc.

UAVs positioning has a major impact on network performance. In fact, the poor placement of UAVs causes a number of issues, such as interference, packet loss, low throughput, and congestion. Thus, an efficient optimization method must be used to address this issue.

UAVs placement belongs to the category of NP-hard problems, successfully solved using meta-heuristic algorithms. In this sense, Shakhatreh et al. [1] suggested PSO for solving the 3D location of a single UAV under disaster situations. PSO was used to minimize the total power needed to serve the indoor customers. Results of simulation demonstrated the effectiveness of PSO when compared to gradient descent (DC) algorithm using various instances. The authors in [2] suggested a hybrid meta-heuristic, called BR-ILS, based on the combination of the

S. M. Taleb (✉) · Y. Meraihi · S. Yahia
LIST Laboratory, University of M'Hamed Bougara Boumerdes, Boumerdes, Algeria
e-mail: s.mekhmoukh@univ-boumerdes.dz

A. Ramdane-Cherif
LISV Laboratory, University of Versailles St-Quentin-en-Yvelines, Velizy, France

A. B. Gabis
Laboratoire LMCS, Ecole Nationale Supérieure d'Informatique, Oued Smar, Alger, Algeria

D. Acheli
LAA Laboratory, University of M'Hamed Bougara Boumerdes, Boumerdes, Algeria

M. D. Hina et al. (eds.), *Future Research Directions in Computational Intelligence*, EAI/Springer Innovations in Communication and Computing,
https://doi.org/10.1007/978-3-031-34459-6_6

iterated local search (ILS) algorithm and the biased randomized (BR) approach, for solving the UAVs placement problem. BR-ILS was validated in terms of the total link capacity and the running time using various instances. Simulation results illustrated the efficiency of BR-ILS when compared to Branch and Cut method and the Naive heuristic. SA was used in [3] for positioning the UAV-BSs. SA was validated in terms of coverage and throughput under distinct scenarios with different numbers of users. In comparison with random placement (RP), results of simulation demonstrated the effectiveness of SA in terms of quality of service and throughput. In addition, authors in [4] suggested a hybrid algorithm, called PSO-Kmeans, based on the merging of Kmeans and PSO algorithms, for addressing the 3D placement of UAVs for emergency scenarios. Signal-to-interference plus noise ratio (SINR) and the total transmit power under various user densities were used to validate the performance of PSO-Kmeans.

Throughput is a crucial indicator of network performance. Throughput refers to information transfer speed measured in bits per second (bps). The throughput of a link depends on the offered load, channel bandwidth, and SINR. The transmission power, the transmitter/receiver distance, and the channel model all affect SINR. Therefore, by adjusting the transmitter/receiver distance, SINR can be controlled. The UAV-based communication system has the opportunity to optimize throughput by positioning the BTS or access point in accordance with user traffic demands. As depicted in Fig. 6.1, we take into consideration a situation where a UAV offers communication services to numerous users who are situated within its transmission range. Some UAV placements will undoubtedly produce higher throughput than others for a given combination of user positions and data rate requirements.

In this chapter, we suggest a hybrid algorithm, called WOA-SA, for solving the UAV placement problem for throughput maximization, depending on data rate demands and user locations. WOA-SA is based on the integration of SA in the original WOA. Using 18 different scenarios, the effectiveness of WOA-SA was

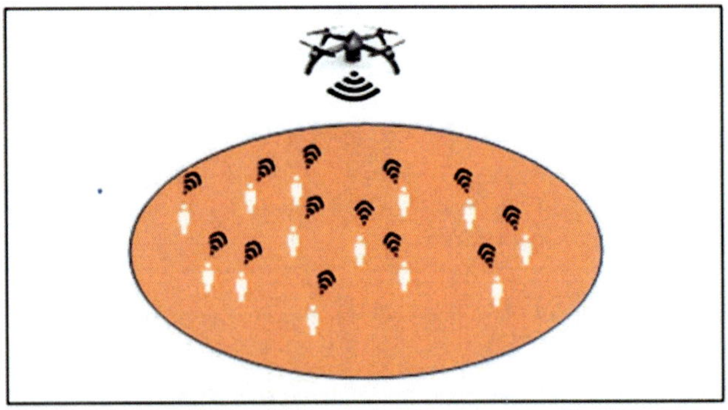

Fig. 6.1 A typical situation. Within its transmission range, a UAV offers communication services to some users. The position of the UAV determines the throughput of each user and the entire system [5]

evaluated in terms of total throughput in comparison to WOA, SA, PSO, GA, and BA algorithms.

The rest of this chapter is structured as follows. Section 6.2 provides a description of the formulation of the UAV deployment problem with throughput maximization. The WOA and SA algorithms are presented in Sect. 6.3. The hybrid WOA-SA algorithm for solving the addressed problem is described in Sect. 6.4. Additionally, we assess WOA-SA effectiveness and present the outcomes in Sect. 6.5. Section 6.6 serves as the paper's final conclusion.

6.2 UAV Placement Problem Formulation

6.2.1 System Model

A three-dimensional deployment region $W \times L \times H$ is used to model the system with:

- A set of ground users S randomly positioned throughout the deployment area. Lgu represents the set of user positions. $Rate$ is the set of traffic demands. All users share the same channel.
- A single UAV u identified by its location Lu, and it is assumed that the UAV flies at constant altitude H and that its coverage radius CR is fixed. The UAV provides communication service for all ground users. The UAV location's total throughput is indicated by the symbol Thr. IEEE 802.11a technology is considered in our work.

6.2.2 Problem Formulation

The main steps for calculating the total throughput for IEEE802.11a technology are given below:

- **Step 1:** The first step consists of determining the physical throughput associated to user j, according to Shannon theorem:

$$Cph(j) = B * Log2(1 + SINR(j)), \tag{6.1}$$

where B is the UAV transmission bandwidth. $SINR(j)$ is the signal to interference and noise ratio at a ground user j. Let $I(j)$ be the total interference at ground user j, $Pr(j)$ represent the received signal power for a link between the UAV and the jth ground user, and N be the noise density. $SINR(j)$ is calculated as follows:

$$SINR(j) = \frac{Pr(j)}{I(j) + N}. \quad (6.2)$$

We assumed that there was no external source of interference to the network. It is crucial to understand that when a MAC protocol is utilized, there is no interference between the users who are connected to the drone. No more than one user can transmit at once on IEEE 802.11 network thanks to the use of RTS/CTS. Thus, $SINR$ is reduced to the signal-to-noise ratio SNR. SNR at a ground user j is calculated as follows:

$$SNR(j) = \frac{Pr(j)}{N * B}, \quad (6.3)$$

where P_t is the transmission power of the UAV, and $G1$ and $G2$ are transmitter and receiver antenna gain. The distance between the UAV and the jth user is represented by $d(j)$. The received signal power is calculated as follows:

$$Pr(j) = Pt * G1 * G2 * \frac{\lambda^2}{4 * \pi * d(j)}. \quad (6.4)$$

The physical capacity of the link between the UAV and jth user is determined as follows:

$$Cph(j) = B * Log2(1 + SNR(j)). \quad (6.5)$$

- **Step 2**: The MAC capacity $C(j)$ and the effective capacity $Ceff(j)$ are determined in this step. In practice, $C(j)$ is less than $Cph(j)$ ($C(j) < Cph(j)$) because of header and routing protocols. However, it is obvious that the higher physical capacity produces higher throughput at the MAC layer and above. $C(j)$ on the standard IEEE 802.11a mainly depends on frame size and physical capacity [5] as shown in Table 6.1. The effective capacity ($Ceff(j)$) is equal to the MAC capacity ($C(j)$) if the jth user does not share the same channel with another user. Otherwise, $Ceff(j)$ is calculated as follows:

$$Ceff(j) = C(j) * t(j), \ \forall j \in S \quad (6.6)$$

$$\sum_{i \in S} t(j) \leq 1. \quad (6.7)$$

Equation (6.7) suggests that only one user can be on the channel at once, essentially eliminating user interference in S. $t(j)$ denotes the airtime of jth user. The max–min fair allocation policy [6] is considered in our work.

- **Step 3**: During this step, the maximum throughput $Thr(j)$ for each user j is determined, depending the data rate demand $Rate(j)$ and the effective capacity $Ceff(j)$ as illustrated in Eq. (6.8). Then, the total throughput of the system Thr

Table 6.1 MAC capacity and CBR packet size for IEEE 802.11a physical rates

Physical capacity $Cph(j)$ Mbps	CBR frame size	MAC capacity $C(j)$ Mbps
54	2264 bytes	33.27
48	2008 bytes	29.59
36	1496 bytes	22.14
24	984 bytes	14.14
18	728 bytes	10.42
12	472 bytes	6.71
9	344 bytes	4.85
6	216 bytes	3.56

associated with the UAV location is calculated according to Eq. (6.9).

$$Thr(j) = min(Rate(j), Ceff(j)) \tag{6.8}$$

$$Thr = \sum_{j \in S} Thr(j). \tag{6.9}$$

The problem tackled in this work consists to determine the best location P_u of the UAV in the deployment area such that the overall capacity Thr is maximized, depending on the locations of users and their traffic demands. The considered issue can be expressed as follows:

$$maximize \sum_{j \in S} min(Rate(j), t * C(j)) \tag{6.10}$$

$$subject\ to \sum_{j \in S} t(j) \leq 1 \tag{6.11}$$

$$Rate(j) - t(j) * C(j) \geq 0, \forall j \in S \tag{6.12}$$

$$(Rate(j) - t(j) * C(j)) * (t(j) - t(k)) \geq 0,\ \forall j \in S,\ \forall k \in S - j \tag{6.13}$$

$$0 \leq x_u \leq W \text{ and } 0 \leq y_u \leq L \tag{6.14}$$

$$d(j) = distance(Lu, Lgu(j)),\ \forall j \in S \tag{6.15}$$

$$C(j) = f(P_t, d(j), I(j)),\ \forall j \in S. \tag{6.16}$$

The objective function (6.10) calculates the total throughput that needs to be optimized. Equation (6.11) guarantees that no more than one person is sending data at a time. Equation (6.12) ensures that a user's airtime allocation does not exceed the user's needs. Equation (6.13) ensures that users are distributed airtime in a fair and equitable manner. Equation (6.14) mandates that the UAV must stay inside the deployment region as it searches for the optimal location. Equation (6.15) calculates the distance between the UAV and the jth ground user. Equation (6.16) illustrates that the throughput capacity is a function of transmit power, the user's distance from the UAV, and interference. In our work, the throughput is computed considering IEEE 802.11a technology as shown above.

6.3 Preliminaries

6.3.1 Whale Optimization Algorithm (WOA)

Mirjalili and Lewis [7] developed the WOA meta-heuristic in 2016. The humpback whales' hunting strategies form the basic concept of WOA. Based on the encircling prey mechanism and spiral updating position approach, the exploitation phase of WOA is simulated. The description is given as follows:

- **Encircling prey: If** $(Pr < 0.5$ **and** $|A| < 1)$
  The position of the solution $W(it+1)$ is updated using Eqs. (6.17) and (6.18):

$$\overrightarrow{D'} = |C\overrightarrow{W}_{best}(it) - \overrightarrow{W}(it)| \tag{6.17}$$

$$\overrightarrow{W}(it+1) = \overrightarrow{W}_{best}(it) - \overrightarrow{A}, \overrightarrow{D'} \tag{6.18}$$

  where it denotes the current iteration, and $\overrightarrow{W_{best}}$ and $\overrightarrow{W}$ represent the best and current solutions, respectively. The vectors of coefficients $\overrightarrow{A}$ and $\overrightarrow{C}$ are calculated as in Eqs. (6.19) and (6.20).

$$\overrightarrow{A} = 2\overrightarrow{a}.\overrightarrow{r1} - \overrightarrow{a} \tag{6.19}$$

$$\overrightarrow{A} = 2\overrightarrow{r}, \tag{6.20}$$

  where $\overrightarrow{a}$ drops linearly from 2 to 0 over iterations (simulating the shrinking encircling behavior as in Eq. (6.19)) and $\overrightarrow{r1}$ is a random vector in range [0, 1]. The formula for a is given below.

$$a = 2(1 - it/t_{max}), \tag{6.21}$$

  where t_{max} is the total number of iterations.
- **Spiral updating position: if** $Pr \geq 0.5$
  The spiral-shaped path followed by the whales is modeled using spiral rule in Eq. (6.22).

$$\overrightarrow{W}(it+1) = \overrightarrow{D}.e^{bl}.\cos(2\pi l) + \overrightarrow{W}_{best}(it) \tag{6.22}$$

$$\overrightarrow{D} = |\overrightarrow{A}.\overrightarrow{W}_{best}(it) - \overrightarrow{W}(it)|, \tag{6.23}$$

where b is a constant, l is a random number in the interval $[-1, 1]$, and $\overrightarrow{D}$ indicates the distance between the current solution and $\overrightarrow{W_{best}}$ at iteration it.

To simulate the exploration phase in WOA (if $Pr < 0.5$ and $|A| \geq 1$), the current whale position is updated by a search agent chosen at random from the population, as shown in Eq. (6.25):

$$\overrightarrow{D} = |\overrightarrow{C}\overrightarrow{W}_{rand}(it) - \overrightarrow{W}(it)| \tag{6.24}$$

$$\overrightarrow{W}(it+1) = \overrightarrow{W}_{rand}(it) - \overrightarrow{A}.\overrightarrow{D}, \tag{6.25}$$

where $\overrightarrow{W}_{rand}$ is a randomly selected search agent from the current population and $\overrightarrow{A}$ is a vector with random values in range $[-1, 1]$.

The main steps of WOA are given in Algorithm 1.

Algorithm 1 The WOA Algorithm

Initialize WOA parameters
Randomly initialize the population of solutions
Evaluate the population and determine X_{best}
for t=1 to t_{max} **do**
  for i=1 to N **do**
    Update a, A, C, l and P
    if $P < 0.5$ **then**
      if $|A| < 1$ **then**
        Calculate $X_i(t)$ using equation 6.18
      else
        Calculate $X_i(t)$ using equation 6.25
      end if
    else
      Calculate $X_i(t)$ using equation 6.22
    end if
  end for
  Evaluate the population and determine W_{best}
  $t = t + 1$
end for
Return the best solution

6.3.2 Simulated Annealing (SA)

In 1983, Kirkpatrick et al. [8] developed the single-based meta-heuristic known as simulated annealing (SA). The main concept of SA is based on the annealing theory that simulates the cooling process of metal atoms. Numerous optimization problems, such as the issue of node placement [9–11], have been addressed using SA.

SA begins with an initial solution X and Temperature Tmp. For each iteration t in $[1tmax]$, SA searches for X' the neighbor of the current solution X. Only two scenarios are acceptable for the solution X': First, if $\delta \leq 0$, where $\delta = f(X') - f(X)$, $f(X')$ and $f(X)$ are fitness values of the neighbor and current solution, respectively. Second, if $\delta > 0$ and the Boltzmann probability $P = e^{\Delta/Tmp}$ is greater than a random value r, the temperature Tmp drops with a cooling factor Cf at the end of the iteration. Up until the maximum number of iterations is reached, this process is repeated. Algorithm 2 illustrates the SA algorithm's pseudocode.

Algorithm 2 The pseudo-code of Simulated Annealing algorithm

Initialize SA parameters: Initial Temperature Tmp_0**, cooling factor** Cf**, and number of the neighborhoods** ns **in the search space.**
Generate initial solution X
while ($t < tmax$) **do**
    while ($it < n$) **do**
        Generate X' using equation
        Calculate $\Delta = f(X') - f(X)$
        Generate a random uniform variable r
        if ($\Delta < 0$) **then**
            Determine X' the neighbor of X
        else
            if ($\exp^{\Delta/Tmp} > r$) **then**
                $X = X'$
            end if
        end if
        $it = it + 1$
    end while
    $t = t + 1$
end while
return The best solution X_{best}

6.4 Hybrid Whale Optimization Algorithm with Simulated Annealing (WOA-SA) for the UAVs Placement Problem

Based on the literature review [12, 13], like most algorithms, WOA suffers from being stuck in local optima and is unable to identify the global optimum solution. To overcome these drawbacks, it is advised that researchers adapt it and integrate it with other strategies or meta-heuristics to improve the exploitation phase. In this sense, we propose a hybrid algorithm called WOA-SA, based on the integration of SA into the original WOA to improve the exploitation phase by searching for the WOA algorithm's most promising regions.

WOA-SA begins by generating initial solutions randomly in the search space. At each iteration, each solution is evaluated using a predefined fitness function. In a comparison stage, if a whale has an improvement in its fitness value in its last iteration, the neighbor solution is determined (SA was applied); otherwise, the algorithm continues to search with WOA. If the stopping criterion is achieved, WOA-SA will be terminated, and the best solution W_{best} will be determined as outputs. The pseudocode of WOA-SA is shown in Algorithm 3.

Algorithm 3 The WOA-SA Algorithm

```
Initialize the parameters of SA and WOA
Randomly initialize the population of solutions
Evaluate the population and determine X_best
for t=1 to t_max do
    for i=1 to N do
        Update a, A, C, l and P
        if (An improvement in the fitness value is done in the last iteration)&(t > 1) then
            Determine the neighbor of the current solution
        else
            if P < 0.5 then
                if |A| < 1 then
                    Calculate X_i(t) using equation 6.18
                else
                    Calculate X_i(t) using equation 6.25
                end if
            else
                Calculate X_i(t) using equation 6.22
            end if
        end if
    end for
    Evaluate the population and determine W_best
    t = t + 1
end for
Return the best solution W_best
```

6.5 Simulation Results

We will evaluate the performance of WOA-SA in this section in comparison to WOA, SA, PSO, GA, and BA algorithms. The algorithms were entirely coded in Matlab. On a computer with a Core i7 2.5 GHz CPU, all simulations are run. We used the Friis free space propagation model because of its simplicity. The total

Table 6.2 Parameters values considered in our simulations

Parameter	Value	Parameter	Value
N° of users n	[8 28]	Coverage radius CR	250 m
Width W	80 m	Transmission power P_t	1 w
Length L	80 m	Frequency f	5 GHz
Height H	20 m	Bandwidth B	20 MHz
Population size N	30	Noise N_0	−101 dbm
N° of iterations t_{max}	2000	N° of runs R	30

Table 6.3 The total throughput under various numbers of users

S	8	12	16	20	24	28
All users are located in sector of 360°						
WOA-SA	**8.45**	**7.41**	**8.16**	**6.19**	4.84	4.57
WOA	7.93	6.82	7.86	5.67	**4.93**	3.82
SA	8.22	6.56	5.86	5.12	4.56	3.86
PSO	8.22	5.58	7.69	5.84	4.68	**4.63**
GA	7.90	6.82	7.86	5.67	4.93	3.82
BA	7.84	5.48	3.96	3.21	3.42	2.82
All users are located in sector of 180°						
WOA-SA	8.09	**8.59**	7.54	5.96	**4.49**	**4.71**
WOA	8.07	7.62	6.44	5.92	4.42	4.52
SA	7.98	6.84	6.95	5.79	4.34	4.46
PSO	**8.23**	8.32	**8.47**	**6.01**	4.47	4.66
GA	5.85	7.55	5.61	4.39	4.21	4.15
BA	8.03	4.44	4.78	4.03	3.98	3.59
All users are located in sector of 90°						
WOA-SA	**16.31**	**12.14**	**11.75**	**9.10**	**9.30**	**7.43**
WOA	14.74	12.14	11.51	9.08	9.92	7.42
SA	16.18	10.85	9.90	7.50	8.36	7.31
PSO	16.31	12.14	11.75	9,10	8.69	7.43
GA	9.84	11.35	8.76	8.23	5.91	7.30
BA	9.51	11.15	6.78	7.21	5.16	7.20

number of iterations is fixed at 2000. A population of 30 solutions was considered in all simulations. Each result shown in this section is an average of 30 trials. The rest of the parameters are shown in Table 6.2.

The evaluation process is made in terms of throughput using 18 scenarios with a different number of users (from 8 to 28), where users distributed in the sector of 360°, 180°, and 90°, respectively. 54 Mbps, 36 Mbps, 18 Mbps, and 9 Mbps are the data rate demands considered in all simulations.

Table 6.3 and Fig. 6.2 depict the results for each of these distributions in terms of throughput under various numbers of ground users. For the three distributions, it is clearly seen that the network performance in terms of throughput decreases as the number of ground users increases. In reality, as the users share the same channel,

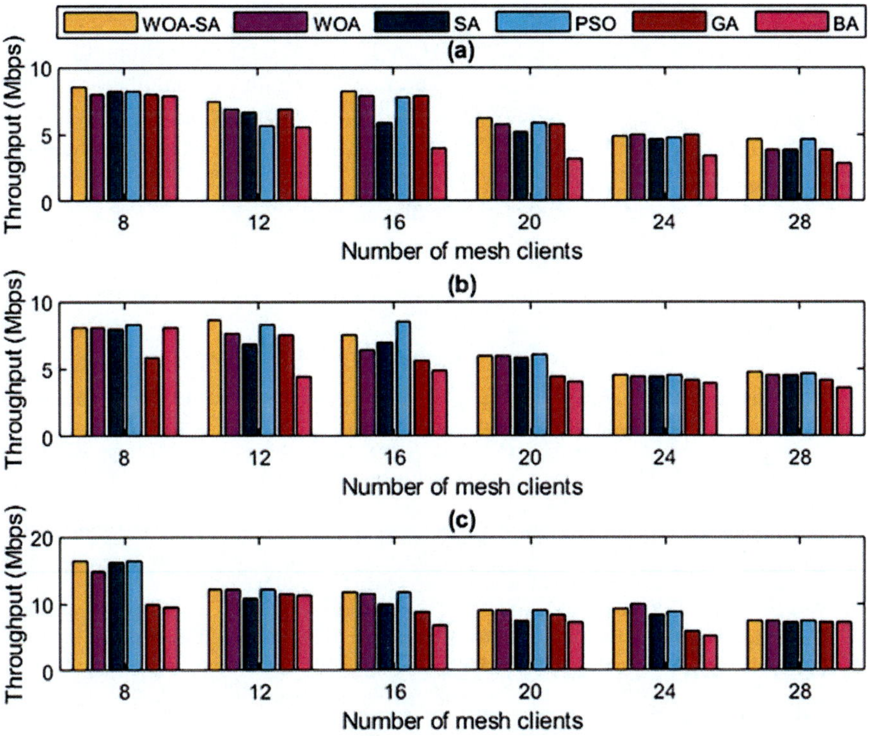

Fig. 6.2 The total throughput under various numbers of ground users distributed in: the sector of 360° (**a**), the sector of 180° (**b**), the sector of 90°(**c**)

the fraction of time or the time user's antenna decreases when the number of users increases, resulting in a decrease in the MAC capacity of each user integrated in the calculation of the total throughput.

In the other hand, for various numbers of ground users, the throughput metric increases as the distribution density of users increases. More specially, the throughput has a major improvement when all users are located in the sector of 90°. In fact, in the case where users are located in 90°, the distance between each user and UAV is reduced, as result the SINR is improved, leading to a major improvement in the system throughput. In addition, for most of cases under different numbers of users and different distributions, it is clearly seen that WOA-SA outperforms in terms of throughput when compared with WOA, SA, PSO, GA, and BA algorithms.

6.6 Conclusion

This chapter proposes a hybrid algorithm, called WOA-SA, for solving the UAV placement problem with system throughput maximization, depending on the user traffic demands and user positions. WOA-SA is based on the incorporation of SA into WOA. WOA-SA was validated in terms of total throughput for three user densities and various numbers of users. The results of simulations using Matlab revealed the efficiency of WOA-SA when compared with five well-known algorithms including WOA, SA, PSO, GA, and BA. For future work, we attempt to use the suggest hybrid approach for positioning multiple UAVs and incorporating the interference caused by sources outside the network.

References

1. H. Shakhatreh, A. Khreishah, A. Alsarhan, I. Khalil, A. Sawalmeh, N.S. Othman, Efficient 3D placement of a UAV using particle swarm optimization, in *2017 8th International Conference on Information and Communication Systems (ICICS)* (IEEE, Piscataway, 2017), pp. 258–263
2. S.A. Fernandez, M.M. Carvalho, D.G. Silva, A hybrid metaheuristic algorithm for the efficient placement of UAVs. Algorithms **13**(12), 323 (2020)
3. N.H.Z. Lim, Y.L. Lee, M.L. Tham, Y.C. Chang, A.G.H. Sim, D. Qin, Coverage optimization for UAV base stations using simulated annealing, in *2021 IEEE 15th Malaysia International Conference on Communication (MICC)* (IEEE, Piscataway, 2021), pp. 43–48
4. W. Liu, G. Niu, Q. Cao, M.-O. Pun, J. Chen, 3-D placement of UAVs based on SIR-measured PSO algorithm, in *2019 IEEE Globecom Workshops (GC Wkshps)* (IEEE, Piscataway, 2019), pp. 1–6
5. Y.-Z. Cho et al., UAV positioning for throughput maximization. EURASIP J. Wirel. Commun. Netw. **2018**(1), 1–15 (2018)
6. E. Danna, S. Mandal, A. Singh, A practical algorithm for balancing the max-min fairness and throughput objectives in traffic engineering, in *2012 Proceedings IEEE INFOCOM* (IEEE, Piscataway, 2012), pp. 846–854
7. S. Mirjalili, A. Lewis, The whale optimization algorithm. Adv. Eng. Softw. **95**, 51–67 (2016)
8. S. Kirkpatrick, C.D. Gelatt Jr, M.P. Vecchi, Optimization by simulated annealing. Science **220**(4598), 671–680 (1983)
9. N.H.Z. Lim, Y.L. Lee, M.L. Tham, Y.C. Chang, A.G.H. Sim, D. Qin, Coverage optimization for UAV base stations using simulated annealing, in *2021 IEEE 15th Malaysia International Conference on Communication (MICC)* (IEEE, Piscataway, 2021), pp. 43–48
10. M. Raissi-Dehkordi, K. Chandrashekar, J.S. Baras, UAV placement for enhanced connectivity in wireless ad-hoc networks. Technical Report (2004)
11. F. Xhafa, A. Barolli, C. Sánchez, L. Barolli, A simulated annealing algorithm for router nodes placement problem in wireless mesh networks. Simul. Modell. Pract. Theory **19**(10), 2276–2284 (2011)
12. F.S. Gharehchopogh, H. Gholizadeh, A comprehensive survey: whale optimization algorithm and its applications. Swarm Evolut. Comput. **48**, 1–24 (2019)
13. H.M. Mohammed, S.U. Umar, T.A. Rashid, A systematic and meta-analysis survey of whale optimization algorithm. Comput. Intell. Neurosci. **2019**, 1–25 (2019)

Chapter 7
Speech Analysis–Synthesis Using Sinusoidal Representations: A Review

Youcef Tabet, Manolo Dulva Hina, and Yassine Meraihi

7.1 Introduction

Speech synthesis is the process of producing artificial speech by computers with the aim to obtain a synthetic speech understandable and indistinguishable from normal human speech [1]. Hence, during the last decades, synthetic speech has been developed steadily in order to improve the intelligibility and naturalness of the speech synthesis system output.

In the context of speech synthesis applications, to maintain good speech quality while minimizing artifacts and naturalness loss, it may be advantageous to encode voice signals using mathematical representations [2, 3]. In order to produce natural sounding speech, a variety of signal processing techniques are being developed for speech synthesis systems [4].

Based on sinusoidal models (SMs) or deterministic and stochastic models (DSMs), several synthesis systems are proposed in the literature [5]. They differ in how the analysis and synthesis are carried out. In the sinusoidal model proposed in [6] for example, a sum of sinusoidal functions evolving over time are used to represent a speech signal. Also, a new approach was introduced by George and Smith [7], based on the hybridization of an overlap-add (OLA) sinusoidal model with an analysis-by-synthesis (ABC). Moreover, to represent in a separate way voiced and unvoiced elements of speech, a harmonic plus noise model (HNM) was suggested in [8, 9].

Y. Tabet (✉) · Y. Meraihi
LIST Laboratory, University of M'Hamed Bougara Boumerdes, Boumerdes, Algeria
e-mail: y.tabet@univ-Boumerdes.dz

M. D. Hina
ECE Paris Engineering School, Paris, France

M. D. Hina et al. (eds.), *Future Research Directions in Computational Intelligence*,
EAI/Springer Innovations in Communication and Computing,
https://doi.org/10.1007/978-3-031-34459-6_7

More recently, adaptive sinusoidal representations (aSRs) have been proposed [10–16]. The adaptive quasi-harmonic model (aQHM), which was proposed in [10], fits the analyzed speech signal onto an exponential basis function with time-varying phase. In contrast, the extended adaptive quasi-harmonic model (eaQHM) [13] also considers the time-varying amplitudes of the basis functions on which the signal is projected. The above aSMs are used to represent only the voiced parts of speech; however, the full-band adaptive harmonic model (aHM) proposed in [14, 15] represents both voiced and unvoiced parts. aHM reconstructs the speech signal as a collection of harmonically linked sinusoids after using the adaptive iterative refinement (AIR) method and the aQHM as an intermediate model to iteratively refine the fundamental frequency. In [16], a novel model known as full-band eaQHM is also proposed being inspired by the full-band aHM, where an adaptivity technique and a continuous basic estimating approach are used. The full-band model assumes an initial harmonic representation that gradually moves toward quasi-harmonicity. It was demonstrated that in order to provide a reconstruction and depiction of high-quality speech, these representations take advantage of the model's local adaptivity on the examined signal.

The rest of this chapter is structured as follows. First, the most popular stationary sinusoidal representations of speech are discussed in Sect. 7.2. Next, in Sect. 7.3, we present the recently proposed adaptive sinusoidal representations of speech. Finally, conclusions and future perspectives are given in Sect. 7.4.

7.2 Stationary Sinusoidal-Based Representations

7.2.1 Sinusoidal-Based Representation

Let $a(t)$ be a time-varying amplitude, K represent a sinusoidal components number, $\omega(t)$ be the instantaneous radial frequency, and $\phi(t)$ be the time-varying phase. The sinusoidal-based representation (SM) suggested in [6] is modeled as follows:

$$x(t) = \sum_{k=1}^{K} a_k(t) \cos(\phi_k(t)) \tag{7.1}$$

with

$$\phi_k(t) = \int_0^t \omega_k(\mu) d\mu \tag{7.2}$$

The model parameters are assumed to be constant during the analysis stage and over short speech frames (local stationarity). A peak picking algorithm based on the short-time Fourier transform (STFT) is then used to compute the sinusoidal components for each frame. Then, a parameter-matching algorithm and a parameter interpolation approach are applied to the sinusoidal parameters of the model. Finally, the reconstructed speech signal can be obtained using Eq. 7.1.

The sinusoidal representation has been well-employed in speech analysis–synthesis systems and has been demonstrated to offer a very accurate reconstruction of speech [6, 17, 18]. However, several models have been suggested to give flexible and excellent representations of speech signals using a combination of sinusoids and a noise term [8, 9].

7.2.2 Harmonic Plus Noise-Based Representation

In the harmonic plus noise (HNM) representation, the speech signal is divided into two elements: a harmonic (periodic) element and a noise (stochastic) element, as the following:

$$x(t) = h(t) + n(t) \tag{7.3}$$

where $h(t)$ is the periodic part defined as follows:

$$ht) = \sum_{k=1}^{K} a_k(t)\cos(\phi_k(t)) \tag{7.4}$$

with

$$\phi_k(t) = \int_0^t 2\pi k f_0(\mu)d\mu \tag{7.5}$$

where f_0 stands for the fundamental frequency; a_k and ϕ_k stand for the harmonic amplitude and phase, respectively.

Let $b(t)$, $h(t)$, and $e(t)$ depict a white Gaussian noise, a normalized all-pole filter with time variation, and an energy envelope, respectively. The noise part $n(t)$ is specified as follows:

$$n(t) = e(t)[h(t, \nu) * b(t)] \tag{7.6}$$

Estimating the harmonic part's properties involves a number of processes. First, initial pitch detection with a voiced/unvoiced decision is computed. Then, a maximum speech frequency is established for each speech frame in order to accurately calculate pitch. The harmonics' amplitudes and phases are obtained by the use of a weighted least squares error criterion. In order to find the parameters of the noise portion, the spectral density of the original signal is represented by an autoregressive filter. After that, this filter uses a triangular variance and temporal envelope to excite a modulated white noise.

In the stage of harmonic synthesis, the phases are unwrapped prior to the application of the linear interpolation, which is done on the amplitudes. The harmonic portion is then readily derived by applying equation 7.4. However, a time envelope ($e(t)$) is used to modulate the output. After that, a white noise ($b(t)$) excites the filter ($h(t)$) for the noise generating stage. Finally, a method of overlap and addition is used to reconstruct the voice signal simultaneously with the pitch.

For a wide range of speech signals, HNM representation has shown to provide good-quality representations. It has also been effectively used to a wide range of speech processing challenges and provides advantages over sinusoidal representation for speech analysis and synthesis [9, 19].

In other hand, both SM and HNM representations add up stationary sinusoids to represent speech frame by frame. A family of adaptive sinusoidal representations was proposed to more accurately represent non-stationary signals, such as speech [10, 13–16].

7.3 Adaptive Sinusoidal-Based Representations

7.3.1 aQHM/eaQHM-Based Representations

In the analysis window, Pantazis et al. [10] presented an adaptive quasi-harmonic model (aQHM) that maps the speech signal onto a collection of time-varying exponential basis functions as follows:

$$x(t) = \left(\sum_{k=-K}^{K} (a_k + tb_k) \exp j(\hat{\phi}_k(t + t_l) - \hat{\phi}_k(t_l)) \right) w(t) \tag{7.7}$$

whith

$$\hat{\phi}_k(t) = 2\pi \int_{t_l}^{t+t_l} f_k(\mu) d\mu \tag{7.8}$$

where t_l stands for the analysis window's center; $w(t)$, $f_k(t)$ represent the frequency trajectory derived using an initial parameter estimation method such as the quasi-harmonic model [20]. $\phi_k(t)$ represents the time-varying phase. a_k represents complex amplitude and b_k a complex slope.

It has been demonstrated that a frequency correction mechanism on the frequency tracks may be used to produce a high-quality quasi-harmonic representation of speech and that its fundamental functions can be altered to suit the input signal's characteristics [10]. Additionally, the hybrid speech analysis–synthesis system termed as the adaptive quasi-harmonic plus noise model has successfully used the aQHM representation [21]. However, only the phase of the adaptive representation discussed above has been modified to account for the speech signal's local features. A novel model known as the extended adaptive quasi-harmonic model (eaQHM) was created in order to incorporate local amplitude adaptation [13].

$$x(t) = \left(\sum_{k=-K}^{K} (a_k + tb_k)\hat{A}_k(t) \exp j\hat{\phi}_k(t) \right) w(t) \tag{7.9}$$

$$\hat{A}_k(t) = \hat{A}_k(t + t_l)/\hat{A}_k(t_l) \tag{7.10}$$

$$\hat{\phi}(t) = \hat{\phi}(t + t_l) - \hat{\phi}(t_l) \tag{7.11}$$

where $\hat{A}_k(t)$ stands for the current amplitude and $\hat{\phi}_k(t)$ denotes the current phase. The eaQHM employs an initialization stage (QHM [20], for example) to estimate the initial values of the amplitude and frequency using linear and spline interpolation, respectively. A frequency integration scheme is used to estimate the phase [10]. A simple least square error criterion is used to estimate the parameters (a_k and b_k). An amplitude–frequency-modulated method, such as [13], is used to repeatedly update the time-varying model parameters (amplitudes and phases). The final synthetic speech signal is reconstructed using the total of these updated time-varying components as follows:

$$\hat{x}(t) = \sum_{k=-K}^{K} \hat{A}_k(t) \exp j\hat{\phi}_k(t) \tag{7.12}$$

The eaQHM representation reaches very precise transcription of speech, better than the aQHM and conventional SM representations, as demonstrated in [13]. Additionally, the eaQHM representation has been effectively applied to simulate a large range of sounds of speech, including unvoiced speech sounds, stop sounds [22], and unvoiced sounds [23].

7.3.2 Full-Band Adaptive Harmonic-Based Representation

Due to the success of the aQHM representation proposed in [10], the authors in [14, 15] suggested to employ a full-band representation to represent both the voiced and unvoiced components as follows:

$$\hat{x}(t) = \sum_{k=-K}^{K} a_k(t) \exp jk\phi_0(t) \tag{7.13}$$

where $\phi_0(t)$ represents a real function defined by

$$\phi_0(t) = \int_0^t 2\pi f_0(\mu) d\mu \tag{7.14}$$

where $a_k(t)$ denotes a complex function that deals with the instantaneous phase and amplitude, and f_0 represents the initial fundamental frequency provided.

Utilizing the specified fundamental frequency curve, Blackman windows are applied to the voice signal at each analysis time. The fundamental frequencies interpolated linearly are carried out using Eq. 7.14 to estimate $\phi_0(t)$.

The aQHM frequency correction technique [10] is used to acquire the model parameters as follows:

$$x(t) = \sum_{k=-K}^{K} (a_k + tb_k) \exp jk\phi_0(t) \tag{7.15}$$

A least squares minimization method is used to estimate the complex parameters a_k and b_k. Using these estimated values, a frequency mismatch error estimate can be calculated, and then the fundamental frequency and the components number can be iteratively updated using a specific adaptive iterative refinement method (AIR algorithm). The current aHM model parameters are constructed by using linear and spline interpolation. Finally, the voice signal is created using Eq. 7.13. Algorithm 1 summarizes the above analysis–synthesis steps [14, 15].

Algorithm 1 Summary of the analysis–synthesis steps

Estimate the complex parameters a_k and b_k of the aQHM
Calculate new frequency mismatch error
Update the frequency and the number of the components using the AIR algorithm
Linear and spline interpolation to obtain the instantaneous aHM model parameters
Reconstruction of the speech signal.

It has been demonstrated in [14, 15] that the FB aHM representation provides high-quality speech reconstruction while dealing effectively with the highly non-stationary character of speech signals.

7.3.3 Full-Band Extended Adaptive Quasi-Harmonic-Based Representation

The eaQHM model in citation [13] has been expanded to a full-band representation employing speech reconstruction and high-quality FB aHM representation [14, 15], where the speech signal is first modeled using an amplitude- and frequency-modulated (AM-FM) full-band decomposition as follows:

$$x(t) = \sum_{k=-K}^{K} A_k(t) \exp j\phi_k(t) \tag{7.16}$$

where $A_k(t)$ and $\phi_k(t)$ denote the current amplitude and phase, respectively, which are determined by the following equation:

$$\phi_k(t) = \phi_k(t_i) + 2\pi/f_s \int_{t_i}^{t} f_k(\mu)d\mu \tag{7.17}$$

For each frame, a fundamental frequency is calculated during the analysis stage. The current amplitudes can then be estimated by assuming a harmonization over the whole band. Then, an adaptive quasi-harmonic representation is derived from the eaQHM representation [13]. Therefore, the signal of speech can be expressed as follows:

$$x(t) = \left(\sum_{k=-K}^{K} (a_k + tb_k)\hat{A}_k(t) \exp j\hat{\phi}_k(t) \right) w(t) \tag{7.18}$$

where $\hat{A}_k(t)$ denotes the estimated amplitude and $\hat{\phi}_k(t)$ represents the estimated phase from the initially harmonic analysis model; $w(t)$ represents the analysis window.

Least squares minimization is used to calculate the complex variables a_k and b_k. A frequency correction term is then obtained using these estimated values. This frequency correction term is used to estimate frequency in an iterative manner, improving the accuracy of the recalculation of the instantaneous amplitudes and phases of the speech signal. Finally, the synthetic speech can be obtained using a sum of its instantaneous parts as

$$\hat{x}(t) = \sum_{k=-K}^{K} \hat{A}_k(t) \exp j\hat{\phi}_k(t) \tag{7.19}$$

Spline and linear interpolation are used to estimate amplitudes and frequencies. To estimate the phase, a non-parametric method is utilized [10]. Algorithm 2 summarizes the above analysis–synthesis steps [16].

Algorithm 2 Summary of the analysis-synthesis steps

Require: Initial fundamental frequency estimation is
Require: Assume a full band harmonicity
Use the eaQHM representation
Use least squares minimization to estimate
Compute a frequency correction term
Estimate Iteratively the frequency
Apply the Linear and Spline interpolation to
Reconstruct the final speech signal .

As demonstrated in [16], the FB eaQHM outperforms alternative speech analysis–synthesis representations such as SM and FB aHM and is a high-quality full-band and voicing-free speech representation.

7.4 Conclusions and Future Work

The overall goal of this chapter was to present a summary of the various methods utilized in speech analysis–synthesis modeling in the literature, giving step-by-step progress in this field. Thus, in this chapter, several sinusoidal analysis–synthesis representations of speech were briefly discussed with main focus on adaptive sinusoidal models.

Compared to the existing state-of-the-art sinusoidal analysis–synthesis representations of speech, the recently suggested adaptive sinusoidal representations of speech showed more practical potential for speech reconstruction and offered high-quality speech.

Future work will concentrate on the design of a multilingual speech analysis–synthesis system based on the adaptive sinusoidal-based models combined with artificial neural-networks-based representations.

References

1. L.R. Rabiner, Applications of voice processing to telecommunications. Proc. IEEE **82**, 199–228 (1994)
2. T. Dutoit, *An Introduction to Text to Speech Synthesis* (Kluwer Academic Publishers, Dordrecht, 1997)
3. P. Taylor, *Text-to-Speech Synthesis* (Cambridge University Press, Cambridge, 2009)
4. Y. Tabet, M. Boughazi, S. Affifi, A tutorial on speech synthesis models. Procedia Comput. Sci. **73**, 48–55 (2015)

5. T.F. Quatieri, *Discrete-Time Speech Signal Processing* (Prentice Hall, Englewood Cliffs, 2002)
6. R.J. McAulay, T.F. Quatieri, Speech analysis/synthesis based on a sinusoidal representation. IEEE Trans. Acoustics Speech Signal Process. **34**, 744–754 (1986)
7. E.B. George, M.J.T. Smith, Speech analysis/synthesis and modification using an analysis-by-synthesis/overlap-add sinusoidal model. IEEE Trans. Speech Audio Process. **5**(5), 389–406 (1997)
8. J. Laroche, Y. Stylianou, E. Moulines, HNM: A simple, efficient harmonic plus noise model for speech, in *Proceedings of IEEE Workshop on Applications of Signal Processing to Audio and Acoustics*, New Paltz, NY, USA (1993), pp. 169-172
9. Y. Stylianou, Applying the harmonic plus noise model in concatenative speech synthesis. IEEE Trans. Speech Audio Process. **9**(1), 21–29 (2001)
10. Y. Pantazis, O. Rosec, Y. Stylianou, Adaptive AM-•FM signal decomposition with application to speech analysis. IEEE Trans. Audio Speech Lang. Process. **19**, 290–300 (2011)
11. Y. Tabet, M. Boughazi, S. Afifi, Speech analysis and synthesis with a refined adaptive sinusoidal representation. Int. J. Speech Technol. **21**(3), 581–588 (2018)
12. Y. Tabet, Speech signal analysis with a refined iterative adaptive method. Int. J. Electron. Commun. Measure. Eng. **11**(1), 1–18 (2022)
13. G.P. Kafentzis, Y. Pantazis, O. Rosec, Y. Stylianou, An extension of the adaptive quasi-harmonic model, in *Proceedings of IEEE International Conference on Acoustic, Speech, and Signal Processing (ICASSP)*, Kyoto (2012)
14. G. Degottex, Y. Stylianou, *A Full-Band Adaptive Harmonic Representation of Speech* (U.S.A, In Interspeech, Portland, Oregon, 2012)
15. G. Degottex, Y. Stylianou, Analysis and synthesis of speech using an adaptive full-band harmonic model. IEEE Trans. Audio Speech Lang. Process. **21**(10), 2085–2095 (2013)
16. G.P. Kafentzis, O. Rosec, Y. Stylianou, Robust full-band adaptive sinusoidal analysis and synthesis of speech, in *Proceedings of IEEE International Conference on Acoustic, Speech, and Signal Processing (ICASSP)* (2014)
17. M.W. Macon, Speech Synthesis Based on Sinusoidal Modeling. PhD Thesis, Georgia Institute of Technology, 1996
18. M. Crespo, P. Velasco, L. Serrano, J. Sardina, On the use of a sinusoidal model for speech synthesis in text to-speech, in *Progress in Speech Synthesis* (Springer, Berlin, 1996), pp. 57–70
19. T. Dutoit, B. Gosselin, On the use of a hybrid harmonic/stochastic model for TTS synthesis-by-concatenation. Speech Commun. **19**, 119–143 (1995)
20. Y. Pantazis, O. Rosec, Y. Stylianou, *On the Properties of a Time-Varying Quasi-Harmonic Model of Speech* (In Interspeech, Brisbane, 2008)
21. Y. Pantazis, G. Tzedakis, O. Rosec, Y. Stylianou, Analysis/synthesis of speech based on an adaptive quasi-harmonic plus noise model, in *2010 IEEE International Conference on Acoustics, Speech and Signal Processing*, Dallas, Texas, USA (2010)
22. G.P. Kafentzis, O. Rosec, Y. Stylianou, On the modeling of voiceless stop sounds of speech using adaptive quasi harmonic models, in *Interspeech*, Portland, Oregon, USA (2013)
23. G.P. Kafentzis, Y. Stylianou, High-resolution sinusoidal modeling of unvoiced speech, in *International Conference on Acoustics, Speech, and Signal Processing*, Shanghai, China (2016)

Chapter 8
Joint Local Reinforcement Learning Agent and Global Drone Cooperation for Collision-Free Lane Change

Jialin Hao, Rola Naja, and Djamal Zeghlache

8.1 Introduction

Lane change is a major cause of vehicle accidents. Therefore, it has attracted attention of road safety stakeholders. Notably, lane change is a real-time critical maneuver that requires special treatment from the driver. In order to perform proper actions, the vehicle driver should watch its surrounding neighbors carefully [1]. Thus, an intelligent and efficient lane change maneuver should be designed for vehicular networks.

On the other hand, wireless vehicular networks are prone to communication gaps with road-side units (RSUs). Moreover, existing lane change assistance (LCA) models focus only on local information such as speed, direction, and distance to the neighbors, which lack a global view of the highway traffic. Thus, unmanned aerial vehicles (UAVs), or drones, can be integrated into the intelligent transportation system in order to overcome communication gaps and to improve the traffic efficiency with their high processing capabilities. In addition, drones can cooperate efficiently with vehicles' on-board units (OBUs) and RSUs owing to their line-of-sight links [2]. Therefore, this chapter proposes an innovative LCA platform coordinated with UAVs.

In the literature, many of the research studies adopt deep reinforcement learning algorithms for the lane change problem. Authors in [3] use a graph-based spatial-temporal convolutional network to predict vehicles' future coordinates and

J. Hao (✉) · D. Zeghlache
Télécom SudParis, Institut Polytechnique de Paris, Palaiseau, France
e-mail: jialin.hao@telecom-sudparis.eu; djamal.zeghlache@telecom-sudparis.eu

R. Naja
ECE Paris Research Center, Paris, France
e-mail: rnaja@ece.fr

M. D. Hina et al. (eds.), *Future Research Directions in Computational Intelligence*, EAI/Springer Innovations in Communication and Computing,
https://doi.org/10.1007/978-3-031-34459-6_8

trajectories. They formulate the best trajectory generation as an optimization problem. However, the study lacks of a real-time performance analysis. In [4], a risk-aware decision-making algorithm based on deep reinforcement learning (DRL) is proposed to achieve the minimum expected risk. Specifically, the risk assessment method uses position uncertainty and distance-based safety metrics. In [5], the vehicle's path for lane keeping or lane changing is represented by a polynomial curve, which is computed from the vehicle's position information: lateral position, longitudinal position, yaw angle, and curvature. Then, the vehicle velocity is generated by applying a quadratic-programming-based optimization model with constraints such as speed limit and safety distance. The deep Q-network (DQN) agent in [6] uses a deep deterministic policy gradient algorithm to make lane change decisions and avoid collisions. In addition, the update delay of the remote vehicle is considered. Performance results show that the agent learns to successfully change lanes with lateral and longitudinal control. Nevertheless, the simulation scenario is based on a single remote vehicle. The authors of [7] propose a DQN-based method that uses a grid-like state representation. The lane change decision is made by a high-level DQN decision-maker, while the velocity is controlled by a rule-based low-level controller. Although 12 interfering vehicles are considered, this study neglects crash performance assessment and safety assurance assessment.

From the previous literature study, we found that a great number of research manage the lane change with vehicle's local information, i.e., speeds, directions, and vehicle positions. However, the global traffic state, i.e., the vehicular density that highly impacts the collision rate, is disregarded. Besides, the reward function does not adapt to the performance results and collision ratios.

Additionally, the rule-based models perform well with pre-defined simulation conditions but tend to fail with unexpected situations, i.e., road with potential risk. Based on this understanding, we propose a LCA platform with DEAR (DEep Q-network with a dynAmic Reward) agent that relies on the DRL approach [8]. In particular, we investigate the model performance under unexpected situations by introducing risky lanes and emergency vehicles that require a higher priority level than other vehicles on the highway.

Consequently, the ego vehicle can intelligently perform a lane change in complex environments. Moreover, the reward function of the proposed agent takes into account the road vehicle density and is dynamically adapted by the drone in real time according to the fluctuating collision ratio. Compared to the literature, our contributions are the following:

1. We propose a lane change decision-making maneuver that jointly integrates global control by drones and local control by DEAR agent, which guarantees a safe and efficient lane change even with the presence of road risks and emergency vehicles.
2. We design two driving modes for the ego vehicle: speed mode and safety mode, depending on the traffic condition around the ego vehicle with the purpose of reducing total travel time while avoiding collisions.

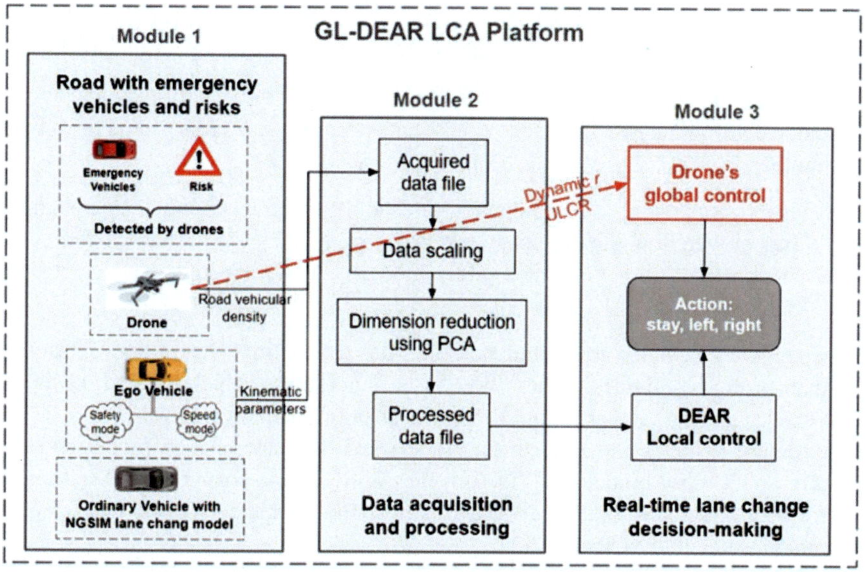

Fig. 8.1 Proposed GL-DEAR LCA platform with three modules

3. We evaluate the performance of the proposed LCA platform, denoted as GL-DEAR with an authentic dataset, NGSIM, generated on a highway in California [9].

The paper is structured as follows: The second section introduces the details of our proposed LCA platform along with its modules. The third section presents an extensive performance analysis. Finally, the conclusion of the paper is provided.

8.2 LCA Platform

8.2.1 LCA Modules

The proposed LCA platform consists of three main modules working in series, as illustrated in Fig. 8.1.

Module 1

Road with Emergency Vehicles and Risks We consider a highway that takes into account potential risks and urgent lane change demands (due to the emergency vehicles). Vehicles on the highway can be divided into three categories: ego vehicle, ordinary vehicles, and emergency vehicles. The proposed platform helps the ego vehicle to make a safe and efficient lane change decision, as shown in Fig. 8.1:

- In our work, the ego vehicle adopts two driving modes: speed mode and safety mode. During the trip, it switches between both modes according to its neighbors number, $n_{neighbor}$:
  - If $n_{neighbor} \leq n_{safe}$ (pre-defined threshold), enter speed mode by assigning higher weight to the efficiency reward.
  - If $n_{neighbor} > n_{safe}$, enter safety mode by assigning higher weight to the safety reward.
- Emergency vehicles are ambulances and police cars that have a higher priority than ordinary vehicles. When there is an emergency vehicle behind, ordinary vehicles and ego vehicle should consider changing lane to give way.
- Ordinary vehicles move according to the NGSIM lane change model and the Krauss mobility model [10]. In fact, the NGSIM lane change model is trained from the authentic dataset NGSIM. The training of the NGSIM lane change model is detailed in Sect. 8.2.2.

On the other hand, drones hover over the highway will gather global information and perform global control to assist ego vehicle's lane change.

Module 2

Data File Acquisition and Processing In the pre-lane change phase, ego vehicle collects kinematic parameters such as GPS coordinates, velocity, and acceleration, while the drones collect the road vehicular density. The acquired data file is then stored and cleaned for a more efficient learning. In this chapter, we apply data processing based on principle component analysis (PCA) and a regression algorithm in order to remove irrelevant and inefficient data. The data processing contains the following steps:

1. Scale the features using Eq. (8.1). In this way, each feature is scaled and translated individually such that it is between zero and one, without breaking the sparsity of the dataset [11].

$$\begin{aligned} x_{std} &= (x - x_{min})/(x_{max} - x_{min}) \\ x_{scaled} &= x_{std} \cdot (x_{max} - x_{min}) + x_{min} \end{aligned} \quad (8.1)$$

2. Reduce feature dimension using PCA. The reduced dimension depends on the number of principle components, n, that we choose. n is chosen by using the *explained variance* v, a measure that maps the variance in the original data to the low-dimensional model, expressed by the eigenvalues λ, i.e.,

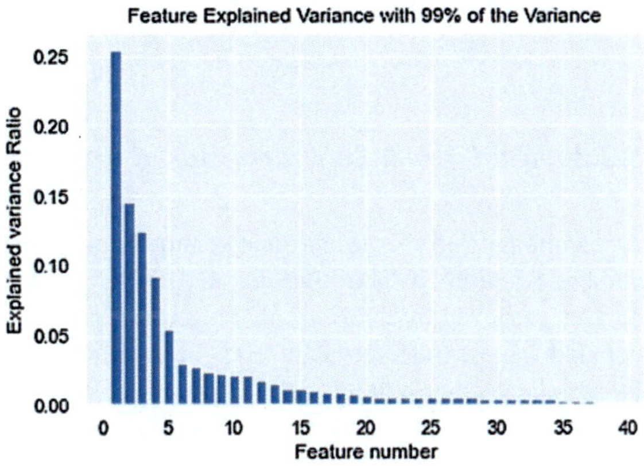

Fig. 8.2 Explained variance ratio of the features with 99% explained variance

$$v_i = \lambda_i / \sum_{j=1}^{n} \lambda_j.$$

Figure 8.2 shows the explained variance ratio of each feature accounted for the whole dataset. It should be noticed that in order to keep 99% of the variance of the dataset, the first 39 components should be chosen; to keep 95% (respectively, 90%) of the explained variance, 28 (resp. 19) components should be chosen [12]. In this chapter, we keep 99% of the explained variance, which leads to a reduced dimension equals 39, in order to achieve the most accurate learning.

Finally, processed data are fed into the real-time lane change decision-making module.

Module 3

Real-Time Lane Change Decision-Making In this module, the drones' global control and the DEAR agent's local control are integrated such that ego vehicle makes a safe and efficient lane change decision in the real time.
The drone's global control is achieved by executing the following steps:

1. At the end of every training epoch, it updates and sends the dynamic collision reward r, together with the road vehicular density to the ego vehicle.
2. As long as the drone detects a road risk ahead (i.e., construction work or car accident), or an emergency vehicle behind the ego vehicle, it sends an urgent LC request (ULCR) to the ego vehicle to force it to change lane. After receiving the ULCR from the drone, the ego vehicle will try to change lane as soon as possible during the valid time of the ULCR, t_{ULCR}.

On the other hand, the local control by DEAR agent is described in Sect. 8.2.3.

The overall algorithm of GL-DEAR platform is described in Algorithm 1:

8.2.2 *Real-World Scenario Based on NGSIM Dataset*

As mentioned before, the lane change model for ordinary vehicles is trained from the authentic dataset NGSIM, thus denoted as NGSIM lane change model. The NGSIM dataset, retrieved by the National Highway Traffic Safety Administration, includes detailed vehicle trajectory data from 4 different neighborhoods in the United States. In this chapter, we adopt the trajectories on a US101 highway to train the NGSIM lane change model. These trajectories are spread over a 6-lane highway without crossings, traffic lights, and pedestrians; the lane 6 is a ramp with vehicles coming

Algorithm 1 `Algorithm for GL-DEAR platform`

Input: E is the number of training epochs, S is the number of steps per epoch, N is the number of ordinary vehicles on the road, t_{ULCR} is the valid time of the ULCR sent by drone, a_g is the global lane change action, a_l is the local lane change action, and a is the final lane change decision
Initialize $t = 0$, $a_g = 0$
for $epoch \leftarrow 0$ to E **do**
    for $step \leftarrow 0$ to S **do**
        for $i \leftarrow 0$ to N **do**
            Ordinary vehicle i predicts its lane change action a_i by NGSIM lane change model
        end for
        Ego vehicle computes $n_{neighbor}$
        if $n_{neighbor} \geq n_{safe}$ **then**
            Enter safety mode
        else
            Enter speed mode
        end if
        Ego vehicle predicts a_l by DEAR ▷ Local control
        if risk or emergency vehicle detected by the drone **then** ▷ Global control
            $t = t + 1$
            if $t \leq t_{ULCR}$ **then**
                $a_g = \{1, 2\}$ according to ego vehicle's current lane
            else
                $t = 0$
                $a_g = 0$
            end if
        end if
        if $a_g == a_l$ **then**
            $a = a_g$
        else
            $a = max(a_g, a_l)$
        end if
        Perform lane change for all vehicles
    end for
end for

in and out from the highway. In order to prevent the exit navigation issue, we investigate the trajectories on lanes 1–5.

For readers' clarity, NGSIM lane change model refers to the authentic model trained from the NGSIM dataset. On the other hand, DEAR is the DQN-based lane change model implemented in the ego vehicle. Both models will be detailed in the following sections. This subsection is dedicated to the lane change model for surrounding vehicles.

Machine Learning Model Applied to NGSIM Dataset

With a comprehensive literature study, we train the NGSIM lane change model based on the eXtreme Gradient Boosting (XGBoost) machine learning model [13]. XGBoost is an ensemble learning algorithm meaning that it combines the results of many models, called base learners (i.e., decision trees) to make a prediction. The most important for training an accurate lane change model from the dataset is to retrieve the most informative parameters. Consequently, we extract 9 features from the NGSIM trajectory data:

$$(v_0, v_1, v_2, v_3, v_4, y_1, y_2, y_3, y_4),$$

where:

- v_0 is the current velocity of the NGSIM vehicle.
- $v_i, i = 1, 2, 3, 4$, refers to the current velocity of the current lane leading vehicle (respectively, the current lane following vehicle, resp., the target lane leading vehicle, resp., the target lane following vehicle).
- $y_i, i = 1, 2, 3, 4$, denotes the distance between the NGSIM vehicle and the 4 neighbors (the leaders and followers in the current and target lanes).

Training and Testing for the NGSIM Lane Change Model

The NGSIM lane change model based on the XGBoost algorithm is trained with the previously obtained training set and labels [13]. We apply the grid-search method to find the optimal parameters for the XGBoost model. As a result, the model with the optimal parameters achieves the testing accuracy of 98%. This model is adopted by ordinary vehicles in later simulations to produce realistic lane change behaviors.

8.2.3 DEAR for the Ego Vehicle

This subsection is dedicated to the DQN lane change agent implemented on the ego vehicle.

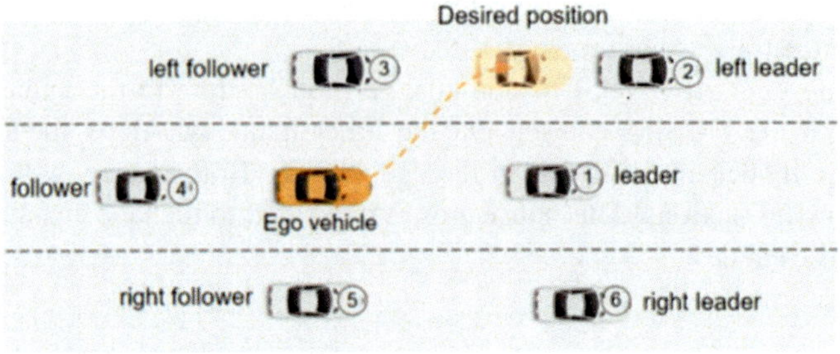

Fig. 8.3 Ego vehicle surrounded by six neighbors, trying to change lane to the desired position

State Space

We train a DQN model of three layers with 64, 128, 64 hidden nodes on the three layers. The output is the lane change decision. The state space at time step j consists of 48 kinematic parameters of ego vehicle and its 6 possible neighbors, namely the leader, left leader, right leader, follower, left follower, and right leader, as illustrated in Fig. 8.3. The observation (or state) at time j is denoted as follows [8]:

$$o[j] = \left\{o_{ego}[j], o_{\Delta}[j], \cdots, o_{\Sigma}[j]\right\}, \text{where:}$$

- $o_{ego}[j] = \left\{l_{risk}[j], x_{ego}[j], y_{ego}[j], v_{ego}[j], a_{ego}[j], l_{ego}[j]\right\}$ is the set of ego vehicle parameters. This set consists of the risk label of current lane l_{risk} (0 refers to no risk detection, 1 refers to risk detection), horizontal position x, vertical position y, longitudinal speed v, acceleration a, and current lane id l at time j.
- $o_i[j] = \{x_i[j], y_i[j], v_i[j], a_i[j], l_i[j], d_i[j], p_i[j]; i \in [0, 6]\}$ consists of 7 parameters of the i-th neighbor, representing the horizontal position x, vertical position y, longitudinal speed v, acceleration a, current lane id l, distance to the ego vehicle d, and vehicle priority p (0 for normal vehicles and 1 for emergency vehicles such as ambulance and police car) at time j.

Action Space

The action space consists of three actions, namely staying in the current lane, changing lane to the right, and changing lane to the left, denoted as $a = \{0, 1, 2\}$.

Reward Function

The reward function is designed with respect of three different perspectives: road safety, travel efficiency, and passengers' comfort. To be more specific, the safety reward is to avoid collisions and potential risks during the trip, the efficiency reward aims to reduce the total travel time, while the comfort reward is to avoid huge changes in accelerations [14]. As a result, the total reward is the weighted sum of the three rewards:

$$R = w_{safe}R_{safe} + w_{eff}R_{eff} + w_{comf}R_{comf}, \tag{8.2}$$

where R_{safe} the safety reward, R_{eff} the efficiency reward, and R_{comf} is the comfort reward. w_{safe}, w_{eff}, and w_{comf} are the weight coefficients:

- Safety reward: The safety reward is the sum of collision reward R_{colli}, vehicular density reward R_{den}, risky reward R_{risk}, and blocking reward R_{block} as follows:

$$R_{safe} = R_{colli} + R_{den} + R_{risk} + R_{block}. \tag{8.3}$$

  The collision reward $R_{colli} = r$ is dynamically adapted by the drone and broadcasted to the vehicles periodically. At the end of an epoch, the drone tunes r according to the following process:

  - Whenever a collision occurs, r is decreased to penalize the collision while discouraging lane change in the following steps.
  - If there is no collision occurs during certain steps, r is increased to encourage lane change in the following steps.

  The density reward R_{den} is defined as the inverse of the vehicle number on the road: $R_{den} = -n_v$, where n_v is the number of vehicles. The reason is that the collision risk is highly related to the vehicular density. The more vehicle on the road, the higher the risk of collision is at lane change, thus lower reward function. It is noteworthy that the traffic density is computed and sent by the drone.

  The risky reward R_{risk} (respectively, the blocking reward R_{block}) is negatively related to the total time of the ego vehicle driving on the risky lane (respectively, in front of an emergency vehicle) such that the agent can learn to adapt to the potential risk on the road and to perform appropriate action with the presence of urgent lane change demand.
- Efficiency reward: The efficiency reward includes the speed reward R_v and the lane change reward R_{change}, as presented in the following equation:

$$R_{eff} = R_v + R_{change}. \tag{8.4}$$

  The speed reward, R_v, is computed as $R_v = -|v_{max} - v|$ so as to get a speed as close as to the maximum speed. The lane change reward, R_{change}, is to avoid frequent and inefficient lane changes, defined as Eq. (8.5), where α, is a constant.

$$R_{change} = \begin{cases} -\alpha, & \text{if a lane change occurs} \\ \alpha, & \text{if stay in lane} \end{cases}. \tag{8.5}$$

- Comfort reward: The comfort reward is negatively correlated to the acceleration fluctuation, as computed by Eq. (8.6).

$$R_{comf} = -\dot{a}_x^2 \tag{8.6}$$

$\dot{a}_x$ is the acceleration difference computed from two adjacent steps.

8.3 Simulation and Performance Results

8.3.1 Simulation Setup

We test the performance of our proposed platform with SUMO [15]. A 4-km circular highway is considered, as shown in Fig. 8.4. The ego vehicle learns to perform a safe and efficient lane change maneuver. Ordinary vehicles adopt the NGSIM lane change model, which has been detailed in Sect. 8.2.2. Four drones are positioned above the highway, communicating with vehicles over vehicle-to-drone (V2D) links, each with a communication range of 1 km [2]. The drones calculate the traffic density in their vicinity by collecting the Hello packet containing kinematic parameters that are broadcasted periodically by vehicles. Additionally, they compute an adaptive parameter r and disseminate it to the ego vehicle over the drone-to-ego (D2E) link.

Vehicle Mobility Model Vehicles adopt the Krauss mobility model with a speed not exceeding 100 km/h. For a more realistic scenario, 10% of the vehicles are "abnormal" with impolite behaviors such as refusing to perform cooperative lane change or driving in an extreme high speed. Furthermore, 20% of the vehicles are emergency vehicles that force in front vehicles to trigger lane change.

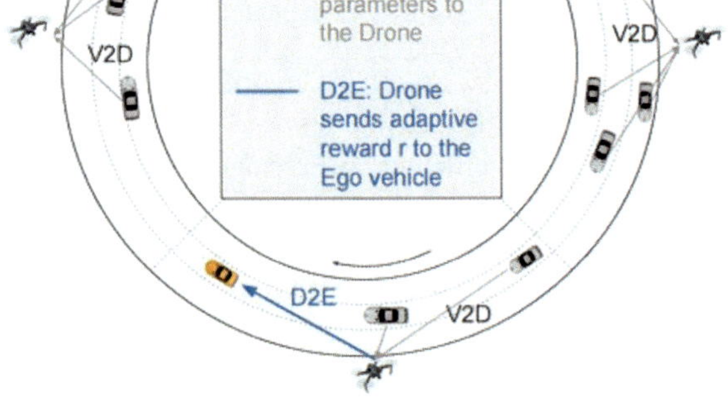

Fig. 8.4 4-km circular highway with V2D and D2E communications. Ego vehicle in yellow, ordinary vehicles in green

Risk Model At the start of each epoch, a random risk, such as a car accident or a construction work in front of the ego vehicle, is detected. The risk is evenly distributed between the three lanes and occurs once at the beginning of each simulation. If a risk is detected on the current lane, the ego vehicle will try to change lanes to prevent from driving on the lane with risks.

The DEAR agent is trained with 150 vehicles on the highway for 10,000 training epochs. Then the trained agent is tested with 10 simulations, each of 100 epochs with different traffic scenarios. We use the traffic control interface (TraCI) of SUMO simulator to retrieve the state information [16]. The agent predicts an action with the state information and computes the corresponding reward according to Eq. (8.2) at each simulation step. Then, the tuple (action, state, reward) will be stored in a replay memory. The detailed performance analysis is provided in the following section.

8.3.2 Performance of GL-DEAR Platform

Baseline Models

We compare our GL-DEAR platform to several baseline models as explained below:

- DEAR: It is a version of GL-DEAR that does not incorporate drone's global control nor the two driving modes of the ego vehicle.
- Policy Gradient: Policy gradient (PG) is a reinforcement learning algorithm. The main idea is to increase the probabilities of high-return actions and decrease the probabilities of low-return actions, until the optimal policy. The reward function for training the PG agent is the same as DEAR.
- K-Nearest Neighbors (KNN): KNN is a supervised machine learning method. It groups data points according to the likelihood, which is estimated by the distance between a new point and a group.
- Deep Neural Network (DNN): A deep neural network is a feed-forward network with hidden layers. The DNN adopted in our comparison is consisted of three hidden layers of 64, 128, 64 hidden nodes.
- Logistic Regression (LR): LR is a supervised machine learning algorithm. It uses a logistic function to model the dependent variables.

The KNN, DNN, and LR models are built with the scikit-learn package in Python [11]. As a result, the testing accuracies of the KNN, DNN, and LR classifiers are 94.2%, 95.85%, and 96%.

Performance Analysis

We compute several performance parameters, namely: collision number, average speed, the number of lane change requests (LC requests), time driving in risky lanes (risky time, t_r), and time spent in front of an emergency vehicle (blocking time, t_b) during simulation. Tables 8.1, 8.2, and 8.3 show the performance of the 6 models tested with sparse (50 vehicles), medium (150 vehicles), and dense (250 vehicles) traffic densities [8].

Collision Number The primary objective for this research is to reduce fatalities caused by accident-related lane changes, which makes the collision number the most important performance parameter. According to Tables 8.1, 8.2, and 8.3, GL-DEAR, PG, KNN, and DNN succeed to avoid collisions in sparse, medium, and dense traffic scenarios. But one can see that KNN and DNN induce much lower average lane change numbers than GL-DEAR in all of the three scenarios. It should be pointed out that a collision-less agent is solely important when lane changes occur. Given this, we found it crucial to investigate the number of LC requests.

Number of Lane Change Requests As a matter of fact, the number of LC requests indicates how differently machine learning and reinforcement learning agents behave. The fact that KNN, DNN, and LR agents achieve fewer LC requests than GL-DEAR, DEAR, and PG implies that KNN, DNN, and LR agents are less likely to allow lane changes, even when there is a risky lane or an emergency vehicle behind. On the contrary, GL-DEAR reinforcement learning agent attempts to learn a safe and efficient lane change maneuver by interacting with the complex environment with its surrounding neighbors, risky lanes, and emergency vehicles, which leads to a high number of lane changes.

Average Speed The most common reason for changing lanes is to overtake a slower vehicle in order to increase speed and reduce total travel time. Thus, we compute the average speed of the ego vehicle during each testing simulation. As it can be noticed from Tables 8.1, 8.2, and 8.3, GL-DEAR achieves higher average speed than KNN and LR in sparse, and medium scenarios. When compared with DEAR and PG, although the average speed of GL-DEAR is lower than the two models, the performance from the safe perspective, i.e., the risky time and blocking time, is greatly improved. The reason is that the frequent change between safety mode and speed mode tunes the trade-off between safety and travel efficiency. The agent learns to increase the speed under the premise of ensuring safety.

Risky Time (t_r) We introduce risky time in order to evaluate the agent's performance with roads prone to risks. According to Tables 8.1, 8.2, and 8.3, GL-DEAR achieves

Table 8.1 Models performance tested with sparse traffic

	GL-DEAR	DEAR	PG	KNN	DNN	LR
Collision number	0	0	0	0	0	2
LC request	604	75	581.3	2.25	0.8	5
Avg speed (km/h)	57.7	76	61.3	56.6	64.9	48.6
t_r (s)	13	46	19.1	82.6	34.9	74.5
t_b (s)	91.2	267.8	191.9	248.6	768.3	0.4

Table 8.2 Models performance tested with medium traffic

	GL-DEAR	DEAR	PG	KNN	DNN	LR
Collision number	0	3	0	0	0	4
LC request	534.5	88	858.6	0.6	1	7
Avg speed (km/h)	54	67.9	61.1	48.7	57.6	46
t_r (s)	8.3	11.1	87.3	51.5	85.4	44.3
t_b (s)	83.2	569.2	332.1	483.4	895.5	71.1

Table 8.3 Models performance tested with dense traffic

	GL-DEAR	DEAR	PG	KNN	DNN	LR
Collision number	0	8	0	0	4	8
LC request	641.6	158.8	749	1.5	22	67
Avg speed (km/h)	49.3	61.1	55.9	50.5	54	42.8
t_r (s)	7.12	74.3	13.2	44	59.5	116
t_b (s)	229.5	492	443.4	620.4	670	5.6

the shortest risky time compared to the other models in all of the three traffic scenarios. In fact, the reward function of GL-DEAR includes a reward related to the risky time. Consequently, the agent will learn to perform timely lane change in the presence of potential road risks.

Blocking Time (t_b) As a matter of fact, the blocking time reflects the driver's cooperation willingness with other vehicles, specifically emergency vehicles. A driver with high cooperation willingness will give way at once when an urgent lane change demand from the following vehicles arises. As shown in Tables 8.1, 8.2, and 8.3, GL-DEAR outperforms DEAR, PG, KNN, and DNN in the three scenarios owing to R_{block} included in the reward function. As a result, the agent can adapt its behavior and gives way when confronted with emergency vehicles, regardless of traffic density.

Impact of Simulation Parameters

In this section, we compare the performance with different t_{ULCR}, the valid time of an urgent lane change request sent by drones to the ego vehicle. Figure 8.5 shows the evaluation of local action and global action at each epoch. One can see that the longer t_{ULCR} is, the stronger the drone's global control over the ego vehicle. According to Table 8.4, the best trade-off between travel efficiency (i.e., average speed) and road safety (i.e. risky time and blocking time) is achieved when t_{ULCR} is 10 s. If t_{ULCR} is too short, it will be difficult to change lane in a timely manner before the ULCR expired. However, if t_{ULCR} is too long, the global control may not be efficient and will even interfere with ego vehicle's local control.

At this point, we can conclude the following:

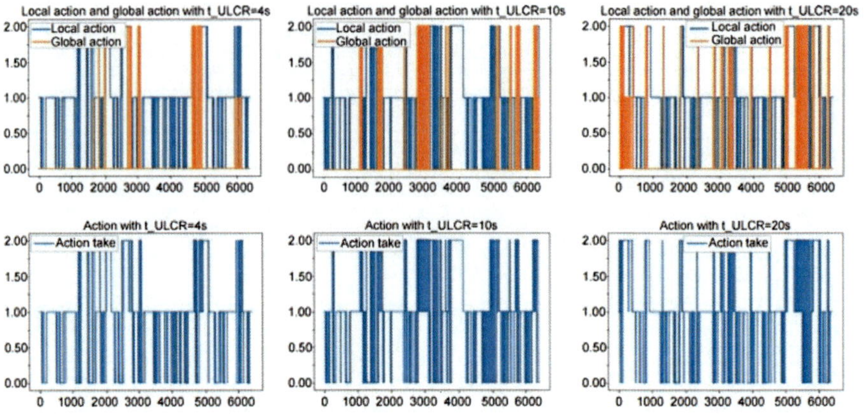

Fig. 8.5 Evaluation of local action, global action, and the action takes with different values of t_{ULCR}

Table 8.4 GL-DEAR performance with different t_{ULCR}

t_{ULCR} (s)	4	10	20
LC request	579.3	534.5	1033.3
Avg speed (km/h)	51.9	54	64.5
Avg reward	1.95	1.98	2.03
Risky time (s)	44.4	8.3	39.6
Blocking time (s)	140.1	83.2	187.5

- The previous analysis proves the enhanced performance of the proposed drone-assisted GL-DEAR platform. Indeed, with drone's global control and DEAR's local control, GL-DEAR successfully avoids collisions while taking into account the road safety, passenger's comfort, and travel efficiency.
- When compared to DEAR and PG agents, the integration of the drones' global control and ego vehicle's two driving modes achieves satisfying performance with GL-DEAR. Although the average speed is decreased, the improvement in safety perspective is huge.
- Whether with sparse, medium, or dense traffic densities, the proposed GL-DEAR agent can learn and adapt to the complex environment despite the presence of potential risks, urgent lane change demands, and emergency vehicles.

8.4 Conclusion and Future Work

In this chapter, we introduce an innovative joint local RL agent and global drone cooperation LCA platform for vehicular networks. Specifically, the drones play a role of collecting global vehicular traffic data and performing global lane change control. The local lane change control is based on the GL-DEAR RL agent with a comprehensive reward function that takes into consideration safety, efficiency, and comfort. The performance is further enhanced by the two driving modes possessed by ego vehicle. In addition, we train a lane change model with the authentic NGSIM dataset in order to evaluate GL-DEAR's performance in a real-world scenario. Performance analysis proves that GL-DEAR achieves collision-less trips and is able to adapt to the complex environment under unexpected situations (i.e., potential risks, urgent lane change demands).

In the next step, we propose to apply federated learning algorithm to meet the real-time requirements of the lane change problem. Drones will work as the central server that aggregates local parameters and trains local models on vehicles in a decentralized manner.

Acknowledgments This work is supported by Labex DigiCosme (project ANR11LABEX0045DI GICOSME) operated by ANR as part of the program *Investissement d'Avenir* Idex ParisSaclay (ANR11IDEX000302).

References

1. J. Bie,, M. Roelofsen, L. Jin, B. van Arem, Lane change and overtaking collisions: Causes and avoidance techniques, in ed. by Naja, R., *Wireless Vehicular Networks for Car Collision Avoidance* (Springer, New York, 2013), pp 143-187. ISBN: 978-14-41995-62-9
2. W. Shi, H. Zhou, J. Li, W. Xu, N. Zhang, X. Shen, Drone assisted vehicular networks: architecture, challenges and opportunities. IEEE Netw **32**(3), 130–137 (2018)
3. Z. Sheng, L. Liu, S. Xue, D. Zhao, M. Jiang, D. Li, A Cooperation-Aware Lane Change Method for Autonomous Vehicles (2022). Preprint arXiv:2201.10746
4. G. Li, Y. Yang, S. Li, X. Qu, N. Lyu, S.E. Li, Decision making of autonomous vehicles in lane change scenarios: deep reinforcement learning approaches with risk awareness. Transp. Res. Part C Emerg. Technol. **134**, 103452 (2022)
5. S. Li, C. Wei, Y. Wang, Combining decision making and trajectory planning for lane changing using deep reinforcement learning. IEEE Trans. Intell. Transport. Syst. **23**(9), 16110–16136 (2022)
6. H. An, J.I. Jung, Decision-making system for lane change using deep reinforcement learning in connected and automated driving. Electronics **8**(5), 543 (2019)
7. J. Wang, Q. Zhang, D. Zhao, Y. Chen, Lane change decision-making through deep reinforcement learning with rule-based constraints, in *2019 International Joint Conference on Neural Networks (IJCNN)*, Budapest (2019), pp. 1–6
8. J. Hao, R. Naja, D. Zeghlache, Drone-assisted lane change maneuver using reinforcement learning with dynamic reward function, in *18th International Conference on Wireless and Mobile Computing, Networking and Communications* (IEEE, Greece, 2022), pp.319–325
9. U.S. Department of Transportation Federal Highway Administration, Next Generation Simulation (NGSIM) Vehicle Trajectories and Supporting Data (2016)
10. S. Krauß, Microscopic modeling of traffic flow: Investigation of collision free vehicle dynamics (1998)
11. F. Pedregosa, G. Varoquaux, A. Gramfort, V. Michel, B. Thirion, O. Grisel, et al., Scikit-learn: machine learning in Python. J Mach Learn Res **12**, 2825–2830 (2011)
12. S. Parisi, S. Ramstedt, J. Peters, Goal-driven dimensionality reduction for reinforcement learning, in *2017 IEEE/RSJ International Conference on Intelligent Robots and Systems (IROS)* (IEEE, Canada, 2017), pp. 4634–4639

13. E. Martínez-Vera, P. Bañuelos-Sánchez, G. Etcheverry, Lane changing model from NGSIM dataset, in *Mexican Conference on Pattern Recognition* (Springer, Berlin, 2022)
14. F. Ye, X. Cheng, P. Wang, C.Y. Chan, J. Zhang, Automated lane change strategy using proximal policy optimization-based deep reinforcement learning, in *IEEE Intelligent Vehicles Symposium (IV)*, IEEE, USA (2020), pp. 1746-1752
15. P.A. Lopez, M. Behrisch, L. Bieker-Walz, J. Erdmann, Y.P. Flötteröd, R. Hilbrich, et al., Microscopic traffic simulation using SUMO, in *21st International Conference on Intelligent Transportation Systems (ITSC)*, IEEE, USA (2018), pp. 2575–2582
16. A. Wegener, M. Piórkowski, M. Raya, H. Hellbrück, S. Fischer, J.P. Hubaux, TraCI: An interface for coupling road traffic and network simulators, in *11th Communications and Networking Simulation Symposium*, ACM, USA (2008), pp. 155–163

Index

M. D. Hina et al. (eds.), *Future Research Directions in Computational Intelligence*,
EAI/Springer Innovations in Communication and Computing,
https://doi.org/10.1007/978-3-031-34459-6